Ornamental Grasses, Bamboos, Rushes & Sedges

Ornamental Grasses, Bamboos, Rushes & Sedges

Nigel J Taylor

WARD LOCK

First published in Great Britain in 1992
by Ward Lock Limited, Villiers House,
41/47 Strand, London WC2N 5JE, England

A Cassell Imprint

Line drawings by Carole Robson

Text filmset in Cheltenham
by Chapterhouse, The Cloisters, Formby
Printed and bound in Spain
by Graficas Reunidas, Madrid

Distributed in the United States by
Sterling Publishing Co., Inc.,
387 Park Avenue South, New York, NY 10016

Distributed in Australia by
Capricorn Link (Australia) Pty Ltd.,
PO Box 665, Lane Cover, NSW 2066

British Library Cataloguing in Publication Data. A catalogue
record for this book is available from the British Library.

ISBN 0 7063 7061 9

Front cover: An autumn view of the grass garden at Pear
Tree House, Litton, Somerset, including *Cortaderia
selloana* 'Pumila', *C. richardii*, *Miscanthus sinensis*
'Zebrinus' and *M.s.* 'Gracillimus' (foreground).

Back cover: The feathery heads of *Stipa calamagrostis*.

Frontispiece: *Agropyron magellanicum* is of lax habit
and has a subtle bluish tinge.

Contents

Preface

Ornamental grasses are enjoying a long overdue surge of interest. The momentum started building in Germany and the United States, where this fascinating group of plants is already an automatic choice for inclusion in both garden and landscape planting schemes. One American company boasts a wholesale list which is in excess of 200 species and varieties, and produces a 30-minute video describing 50 of these. Elsewhere, there have been voices in the wilderness, but it has to be said that the ball has only recently started to roll in earnest. Thankfully it is now well and truly on its way, as the frequent coverage on television gardening programmes and in the gardening press testifies, together with increasing availability from garden centres and nurseries.

Any relatively new interest is inevitably accompanied not only by curiosity, but by mystery, misconception and even prejudice, and some encouragement, reassurance and advice can be most welcome. That is the purpose of this book. It aims to bring the grasses and their allies down to earth, to allay commonly expressed fears, and to whet the appetite of the reader for a closer, first-hand relationship with a group of plants whose unique features can breathe fresh life into his, or her, garden.

Hard architectural lines, and the precision and unyielding rigidity of man-made building and paving materials need softening. Gardeners and landscape designers recognize the value of foliage in achieving

Agrostis canina **'Silver Needles' at its flowering peak, seen here with the cream buds and silver foliage of the curry plant,** *Helichrysum serotinum.*

this, but surely no other group of plants can match the grace of line and curve inherent in the foliage of grasses, nor the essential delicacy and airiness of so many of their flower heads. And is there another group so mobile, so ready with an animated response to the slightest breath of wind? Add to this an amazing variety of foliage colour, and general ease of cultivation, and you have a large range of plants, virtually unexploited, yet increasingly available, just waiting to contribute all these exciting qualities to the search for peace and pleasure in the garden.

A significant proportion of the grasses, sedges and rushes that are available to the gardener, together with a representative selection of the large number of bamboos, some 240 plants in all, are discussed, with information for their successful cultivation.

Unfamiliar botanical terminology has been carefully avoided in an effort to convey important information in an acceptable way to today's gardener whose life is already sufficiently full of complexities, and who is usually seeking mental and spiritual relief, not further complication, from gardening.

Strictly speaking the term 'grasses' correctly includes bamboos, but not the sedges or rushes, which are quite different botanical families. They are normally considered together because, from the point of view of superficial appearance and garden use, they share a number of characteristics. In the interests of conciseness the word 'grasses' is here used to embrace all of the above groups, and in most of its occurrences it carries this broader sense.

N. J. T.

7

CHAPTER 1

The Truth About Grasses

There is a wry smile on the face of the visitor to the flower show as she contemplates the display of ornamental grasses, carefully arranged in their pots on the stand. She was heading for the fuchsias when a chance sideways glance checked her progress towards those enticingly bright colours. Intrigued, she approaches the display, but is clearly hesitant about coming too close. Slowly she starts to circle the stand, though keeping at a safe distance.

The exhibitor has noticed the lady, and observes her cautious approach. He has seen the same thing many times before. He watches as she is drawn, almost involuntarily it seems, around the stand, still wary, but now venturing closer, clearly attracted, even if reluctantly, by what she beholds. She completes her tour. Now, with a knowing expression on her face, she approaches the perpetrator of this display. He knows just what is coming: 'Quite nice, aren't they?' she comments, studiedly understating her true impressions, and then, in a tone that might equally as well have been directed at an importer of cute, but rabies-carrying puppies, 'But, of course, you can't let them loose in the garden!'

This oft-repeated experience characterizes the foremost cause of prejudice against the grasses as garden-worthy plants. 'Grass gets everywhere', our sceptic reasons and, when you think about it, there is no doubting the truth of that statement. Grass *does* get everywhere – or almost – often reaching (to borrow a phrase) the parts that other plants cannot reach! The grasses, along with the sedges and rushes, some 14,500 species, are among the most successful plant types in the world, and are surely the most evident. Recall the last time you took a summer walk in the country. You may have noticed many flowers in the hedgerows, clothing the banks and roadside verges, but you can surely not have failed to observe that these were in the minority compared with the abudance of narrow leaf blades and airy or plume-like flower heads of the grass family. Meadows are a delight at certain times of the year when they are studded with the bright colours of a great variety of wild flowers. But what is the background against which these gems are displayed? And what remains when the flowers are spent, if it is not a veritable carpet of wild grasses? Woodland edge, forest glades, moorland, coastal sand dunes, ponds, lakes and their margins, indeed every ecological ingredient of our varied landscape boasts its own grasses and grass-like plants, and almost invariably they constitute the predominant vegetation. Prairie, plain and steppe afford vistas of little else but grasses. The vast areas of land that man has cultivated he has put down to grasses more than to any other crops. Corn, rice and maize, to name but three staples, are grasses, and among their further diverse types are numerous basic foods for both man and animal, besides others which are utilized commercially in a multiplicity of ways. When it comes to our sports fields, parks and, of course, gardens there is no more versatile, practical, relatively easily maintained – and pleasing – surface than the grass sward.

Yes, there is no denying the fact – grass does get everywhere! The question is, should we allow this truth to deter us from introducing grasses into the garden for any purpose other than as a lawn? Before answering this question let us allow our contender to voice a further doubt that is worrying her: 'Grasses are...' she hesitates, '...well, different, aren't they? I just wouldn't know where to start with them. Where do I put them? What do they like?'

The suspicion that the grasses and grass-like plants are 'different' is botanically true in quite a number of rather technical respects, but the cause of the trouble for the average gardener seems simply to be the visual perception that they have neither 'ordinary' leaves nor 'normal' flowers. Strangely, this superficial dissimilarity somehow appears to separate them by light years from the traditionally available range of familiar, 'ordinary' plants. Quite unaccountably it seems to remove them from the realm of reason which normally moves us unquestioningly to apply basic principles in our cultivation of all other commonly accepted, though still widely differing, plant types.

If either, or both, of these considerations have inhibited you from using ornamental grasses in your garden, please reflect for a moment on this reassuring statement of fact: from a gardening point of view perennial grasses may be equated exactly and absolutely with any other perennial garden plant. A like correlation may be made between annual grasses and any other annual garden plant. This basic premise immediately brings the grasses down to a level at which we can feel entirely comfortable with them, because the only difference *in gardening practice* between them and our 'normal' garden favourites is a visual one – and that is a positive blessing because, as we shall see, it gives us unique features that we may exploit to our benefit. A further happy implication is that there is a grass, just as there is a herbaceous perennial, for every spot. Indeed there are conditions where a grass is the only plant that will suit. Whatever the nature of your garden then, there will certainly be a good choice of grasses and grass-like plants from which to select that will be perfectly content with whatever you have to offer.

In the hope that at least some of the aura of mystery that often seems to shroud the subject has been dispelled, let us return to the first objection – that grasses get everywhere. Agreeing that they certainly appear to do so, our still decidedly dubious friend has an idea as to why that is the case. 'Don't they seed themselves prolifically?' she asks. 'After all, half of the weed problem in my borders is grass seedlings.'

Indeed, there are some prolific self-seeders. Many of these, however, are among the wild rather than the cultivated grasses. Even so, there are one or two ornamental species and varieties with similar reproductive habits. But then are there not other hardy perennials in our gardens whose seedlings we seem to find everywhere, and are constantly removing? If that really annoys us we throw the plant out, or perhaps avoid it in the first place. However, more than likely it makes a contribution to our planting scheme that we appreciate, and we are prepared either to tolerate its little excesses or to remove its seed heads before any damage is done. Such free self-seeders are in a tiny minority and, please note, this fact is as true of the grasses as it is of their fellow hardy perennial companions.

'Fair enough,' the visitor is gradually coming round, but she hasn't finished yet. 'I can grasp the idea that they are just like ordinary perennials but, even if they don't seed themselves everywhere, surely they run around and swamp everything within sight.' Again, yes, just a very few do. But again, so do some of the other plants in our gardens. That may well be the very reason for our planting them. We need some plants that will perform thus. We used them in full awareness of their habit of growth, and because they would do just what we wanted them to – that was to cover the ground. Nothing else would have done the job. We place them deliberately where they will fulfil their role and avoid planting them where they would overrun other treasures. We acknowledge their rampant nature and if they are a little over-exuberant we restrain them once or twice a year by removing the bits that have exceeded their bounds. Again though, let it be emphasized, the number of grasses that are of running habit is very small. By far the majority, as with other hardy perennials, are clump forming.

The lady smiles, thanks the exhibitor for his patient explanation and reassurance, pays for a catalogue with the promise that she might just try a grass or two, and proceeds once more in the direction of the fuchsias. The diversion, she feels, was not a waste of time.

Some basic botanical information

The purpose of this book is to consider the garden uses of the grasses, bamboos (essentially woody-stemmed grasses), sedges and rushes. Technical descriptions of the botanical differences within these families, and between them and other plant groups are available in other works, and only the briefest consideration seems appropriate here.

The visible parts of the plant are the stems, leaves and flowers. In the true grasses, including bamboos (Gramineae), and sedges (Cyperaceae) the stems are called culms. If you can lay your hands on a bamboo cane in the garden shed a quick look at it will reveal the three basic features which are, with very few exceptions, common to the whole Gramineae family: the culms are cylindrical, they have swollen joints called nodes, and they are hollow (except at the nodes). The sedges are different, and may therefore be readily distinguished, in that the culms are triangular in cross-section, they have no nodes, and are pith-filled, therefore solid. The rushes (Juncaceae) are generally cylindrical like the grasses but, in common with the sedges, are solid and nodeless.

The leaves are, of course, characteristic, being narrow and linear (though surprisingly broad in some cases). The rushes and sedges share an arrangement where the leaves progress in three ranks up the shoots, whereas in grasses they alternate in two opposite ranks up the culms.

So far as foliage is concerned we do find grassy and sword-like leaves in quite a number of other plants, but an examination of the flowers of the grasses reveals that they are in a world of their own – a world, not of brightly coloured invitations to buzzing bees to alight and unwittingly collect and deposit the pollen that will ensure the successful production of seed, but of wind pollination where minute pollen grains in their inconceivable millions are wafted between neighbouring or more widely separated plants. Whilst hay fever sufferers may have good reason to rue this phenomenon it none the less allows for a quite distinctive assortment of flower heads of unrivalled grace and subtlety. The true flowers are quite tiny and only in the case of the rushes do they in any way resemble 'normal' flowers. Usually a number of these tiny flowers combine in what is termed a spikelet, and in turn the full complement of spikelets constitutes the flower head, or inflorescence. There are three modes of arrangement of the spikelets which contribute to the appearance of the flower head, each with its descriptive term (Fig. 1): a spike is an inflorescence whose spikelets attach directly to its main stem (axis) without stalks. They will usually, therefore, be fairly tight and narrow. Sometimes the spikelets grow on short stalks, themselves attached to the axis. This is a

The dense hairs of *Alopecurus lanatus* give it a silvery effect. Chippings around the crown discourage rotting in winter wet.

10

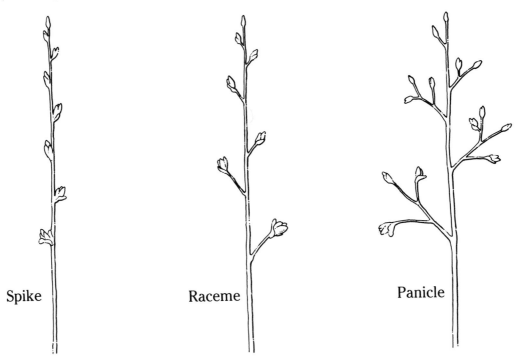

Spike   Raceme   Panicle

Fig. 1 The three main types of inflorescence.

raceme and will again tend to be quite narrow. The glorious plumes and open heads of many of the grasses are called panicles, where the spikelets are carried on stalks which themselves are on branches from the main stem.

As far as their aesthetic contribution to the garden is concerned grasses, bamboos, sedges and rushes all feature the characteristic, essentially erect or arching form.

The true grasses offer the greatest wealth of diverse floral beauty. There are tremendous variations in plant size and form, and in foliage size, arrangement and texture. Foliage colour is generally green (sometimes evergreen), with a small number of yellows, blues and reddish purples, and variegated forms are well represented.

The bamboos offer nothing from a floral point of view, and foliage colour is limited to shades of green although they do include some fine variegated forms and are almost invariably evergreen. They range in size from small, though often with extensive ground-covering capacity (i.e. invasive!), to very tall and almost tree-like, with a superlative grace of form.

For a kaleidoscope of fascinating foliage colours, often evergreen, the sedges are surely unrivalled, and this more than compensates for generally insignificant flower spikes and, with exceptions, limited ranges of form and size.

The rushes, including woodrushes, offer just one or two exciting, and two or three interesting forms, the remainder featuring only modest attractions, but being useful in more extreme conditions of wet or dry.

CHAPTER 2

Why Grow Grasses?

So far we have endeavoured to allay any immediate fears, and we have at least considered why, to use a double negative, you shouldn't *not* plant grasses! But you may feel that you are quite happy with your garden as it is, using, as you are, an already good variety of plant types. Why bother with having to get to know yet another group with its quirks and idiosyncrasies? Is the choice not already overwhelming without introducing more elements? Let us embark upon the positive aspects of the matter; in other words, why grasses are a must for the garden.

As gardeners desirous of using our plants much as an artist would his paints, we have available to us a varied and extensive palette of different plant types: trees, shrubs, conifers, heathers, alpines, ferns, herbaceous perennials, aquatics – and grasses. Each group is distinctive, unique in certain respects. Each offers its own peculiar features, contributing characteristics with which it alone has been endowed by the great Creator, and which no other group can anywhere near duplicate. In achieving maximum interest in our planting surely we should be actively seeking to include respresentatives of each group, each bestowing its own singular traits, thus availing ourselves of a little bit of everything that the wonderful plant world has to offer. Conversely, there would seem to be little point in depriving ourselves of something that could only enhance the overall effect. Now when it comes to the grasses we have a group of plants which is so individual, so distinct in more than one respect, that it absolutely cannot be excluded – it just must be represented somewhere. Just what is it, then, about the grasses that makes them so worthy of inclusion in the garden?

Form

First and foremost consider their form – simple and vertically linear. There is nothing fussy about any of the grasses and they are, in consequence, intrinsically restful. Their ubiquitous distribution throughout the earth's ecosystems means that, when introduced into the garden, they inevitably bring with them something of the ethos of their natural habitat. The impact is both visual and auditory, especially in the case of the medium and larger grasses, evoking the sights and sounds of wide open spaces, a mysterious synthesis of the wild and the serene. Meadows and barley fields are conjured up, rippling and gently swishing in the breeze; the vast panorama of American prairie grasses constantly in motion like rolling ocean waves; or, in a different setting, the rustling of reed beds on wide, wild, windswept marshes. Surely no other representatives of the plant world can so easily distil the very essence of those parts of our earth that are the furthest removed from the troubles and cares with which mankind seems to have burdened himself. How refreshing, then, to be able to indulge in a little romantic relief simply by stepping out into one's garden.

There are a number of plants with more or less linear leaves, but the great majority of plants' leaves are very definitely broad and rounded in general outline. Likewise the overall shape of practically all other types of plant is decidedly rounded. They sit so densely solid, stolid, immovable – unless a real wind gets hold of them. By contrast, observe the grasses – always the first in motion in the slightest breeze. The repetition of line and curve affords a total change of texture, an important ingredient in the recipe for successful planting. They stand almost alone in giving a marvellous vertical lift, an essentially upward thrust that is so refreshingly different that it offers a wonderful contrast to our other plants. That contrast must be exploited. In current parlance they are the yuppies of the plant world – upwardly mobile in a way few other plants are.

Foliage colour

What a wonderful gift colour vision is. We could get by in monochrome (remember black and white television?), but colour immeasurably enriches life's experiences, and it seems to be the factor that makes the most immediate impression on our senses. There is no end to the array of bright colours in the plant world, largely represented in their flowers. A group of plants that is rarely anything more than subtle through virtually its entire range is therefore of great value, again by way of contrast. That group is, of course, the grasses, and it is foliage rather than flower colour to which we are here referring. Very few strong colours are reckoned among their number. *Carex elata* 'Aurea' (Bowles' golden sedge) and the bamboo *Pleioblastus viridistriatus* are a strong yellow; *Milium effusum* 'Aureum' (Bowles' golden grass) only slightly less so; *Uncinia rubra* is a rich foxy red-brown; *Imperata cylindrica* 'Rubra' a startling blood red, and the more strongly white-variegated varieties are certainly bright. Possibly the almost electric blue *Agropyron magellanicum* could be included too. But such are exceptions and are few and far between.

The point must now quickly be made that subtlety should not be confused with dullness. In fact the range of hues to be discovered among the grasses is nothing short of exciting. We have just mentioned the brighter yellows, reds, browns and blues, and the white-variegated forms. There are also yellow, cream and tri-coloured (i.e. white, pink and green) variegations, generally in the form of longitudinal stripes, but with yellow or cream cross-banding in a handful of intriguing cases. The sedges contribute some further marvellous colours, several of them quite unmatched by any other plants: silvery and whitish greens, pale and deeper olives sometimes overlaid with bronze to orange overtones, grey- and chocolate-browns through to red- and rich orange-browns, and even maroon. There are two or three variegated rushes; most are green, but there is one, the woodrush *Luzula ulophylla*, whose hairy backing to concave leaves gives a very silvery effect. Back with the grasses there is a woolly silver (*Alopecurus lanatus*), beautiful silvery blues, powder blues, purplish blues, grey-blues; blue-greens, grey-greens and mid-greens; yellows, ochres and fresh yellow-greens; deep reddish and bronzy purples and the strong red of *Imperata* just mentioned. Here, then, is a second factor which we may advantageously turn to our account: an incredible palette, predominantly subtle, a marvellous addition, and foil, to the standard plant colour range.

Flower power

A third invaluable contribution by the grasses to the garden scene is furnished by their flower and seed heads. While retaining the vital reproductive organs of all plants they bear absolutely no resemblance to 'ordinary' flowers, as commented earlier. Adjectives such as airy, open, graceful, fluffy, lacy, shimmering,

Alopecurus pratensis 'Aureovariegatus' with the shrub Physocarpus opulifolius 'Dart's Gold' and Acaena inermis 'Copper Carpet': totally different leaf forms in a pleasing association.

feathery, diaphanous, wispy and silky, and nouns like plumes, spikes, sprays, clouds, and even caterpillars, shuttlecocks and bottle-brushes could rarely be applied to 'normal' flowers. One can appreciate, then, how the word texture may frequently be appropriate with reference to the flowers of grasses, whereas it is scarcely applicable otherwise in floral terms. Yes, they are different. Now, difference means contrast, and contrast is the breath of life to the garden.

Perhaps the best known among the inflorescences (flower heads) of the grasses are the glorious shaggy plumes of the better forms of pampas grass. But there is tremendous variation, ranging from short stubby spikes to the most delicate, light and airy open panicles. The individual spikelets also manifest great variety, some resembling the tiniest beads, others larger, like those of the quaking grasses (*Briza* spp.), heart-shaped and nodding, dancing in the slightest breath of wind. (Indeed the response of their flower heads to the wind is one of the most appealing characteristics of the grasses as a family.) Many feature hairs or bristles to a greater or lesser degree, such as the pampas grass just referred to, or the cotton grasses (*Eriophorum* spp.). Cultivated barley and members of the *Stipa* family are further examples. In all cases the colouring is subdued but, none the less, as with the foliage, manifests a fascinating diversity, including quiet shades of green, yellow, blue, violet, purple, brown etc. The close and comparative examination of the inflorescences of several different grasses is a delightful exercise. Generally speaking, while there are some decorative seed heads among other 'ordinary' flowering plants, many lose their appeal totally at this stage, whereas the grasses tend to hold their attractive appearance for a longer duration, right through the whole flowering and seed-forming process.

Ease of cultivation

Always a factor to consider when selecting plants is ease, or otherwise, of cultivation, and this must rank as a further plus in our contemplation of the merits of the grasses. Remember that they may be regarded simply as straightforward perennials or annuals. The vast majority of such plants are perfectly happy in, or at the very least will uncomplainingly tolerate, most average garden soils and situations. A few will sulk if it is too dry or too wet, too sunny or too shady, too cold, too hot, or too windy. The same is true of grasses. In general they are obligingly long suffering and only too anxious to please. The range of conditions that some grasses will tolerate is actually quite incredible. Some that are normally associated with average soil conditions will actually grow in a garden pool. Others that grow naturally in water show no signs of resentment at being planted in average soil in good light. At the other extreme are those whose native habitat is sun-baked and dry, and yet which will likewise accept not unduly free-draining soil in good light with good grace. Perhaps the point that is being made is that if you have 'average soil and good light' there is not much you cannot get away with. And if you are prepared either to improve the drainage or to add moisture-retentive organic matter as appropriate to a particular species, most of the remainder may be accommodated quite happily. The odd few that might let you know if you offend them by siting them incorrectly are mentioned in the alphabetical list (Chapter 5).

CHAPTER 3

Using Grasses in the Garden

Broad principles of contrast and colour

The word contrast has appeared several times already with the implication that it is a feature of no little significance in garden planning. Indeed we could perhaps go so far as to say that in devising our planting schemes contrast is the single most important factor for which we should be striving. Above all else it breathes life into a scene. It creates interest.

In the many different types of plant available to us we have all the ingredients of an interesting scene. The more of those different types of which we avail ourselves the more life we inject. You, as the gardener, face the challenge of arranging the material at your disposal in the most aesthetically pleasing manner possible. This challenge is at once both daunting and exhilarating, one with which you will most likely always find yourself grappling in the search for total satisfaction with the result that you have produced. Thankfully there is great scope for contrast using any of the following factors, either alone or, much more likely, in combination.

1. COLOUR Put a blue-leaved plant against a yellow one and you have contrast.
2. TONE Plant dark green leaves next to paler green and again you have contrast, or, associating the same dark green with silver would give a double, and therefore stronger and more lively contrast – of colour and of tone (lightness/darkness).
3. FORM Place erect against prostrate – another contrast – or dome-shaped with arching.
4. HABIT Clump-formers contrast with carpeters.

5. HEIGHT Contrast low with tall. Among the grasses there are miniatures for rock gardens or sinks, with bamboos many metres high at the other extreme.
6. FOLIAGE Apart from colour and tone a wealth of possible contrasts are available, even among the largely linear grasses: narrow with broader, shiny with matt, smooth with hairy, large with small, long with short.
7. FLOWERS Again contrasts galore are available, in shape, size, colour etc. However, remember that flowers are generally a decidedly more temporary feature and that other forms of contrast should be sought for the rest of the year. Where the distinctive inflorescences of the grasses are concerned we may enjoy the contrast between long and short, or between those which are tight, dense or narrow, and the more open, broad and airy examples.
8. DECIDUOUS/EVERGREEN Do not forget that in winter the contrast between evergreen and deciduous or herbaceous plants will be appreciated, and the former should be borne very much in mind when devising planting schemes.
9. MASS The relative number of two or more types of plant in itself provides contrast, besides balance. For example, several low rounded plants may be nicely balanced by just one contrasting vertical grass.

These nine sections embrace many times that number of contrasting permutations, and there is tremendous scope for ingenuity. Indeed it would be a fair-sized garden that was large enough to accommodate anywhere near the full range of possibilities.

The point made in section 9 is an important one. We will usually wish to avoid a too-busy, fussy effect. A

degree of restfulness is what most of us cherish in a garden, and to that end some repetition, some continuity, is appropriate. Actually the grass lawn serves that purpose quite admirably. It is intrinsically simple, a flat monotypic, monochromatic surface – the perfectly serene foil for the animation of the border. But a similar effect may be achieved within the border itself by planting several of the same, or similar plants together. This is no doubt why we are usually exhorted to plant in groups of three, five or seven (which in itself is intriguing, the uneven numbers indicating, as they seem to, our innate antipathy towards the symmetrical and formal when dealing with the things of nature). Having attained the required serenity, some kind of balance must now be sought to avoid monotony – hence, in the example above, the one vertical accent complementing the repose of the adjoining group. An area of more gently coloured plants may need balancing with one smaller splash of brightness. The reverse is also true. Any patch of vivid colour will benefit from the tranquillity of a larger balancing area of quieter hues. Each enhances the other, as well as the scene as a whole.

Reference has already been made twice to colour: first with regard to the range available among the grasses, and second as a source of contrast. To develop the subject a little further, contrast is not the only use to which we can put colour. We can contemplate and enjoy a colour quite alone and in its own right, usually when observing a plant at close quarters without reference to its neighbours. The moment we step back a bit though, and the neighbours assert themselves, we start to see colours in relation to one another. The

The flower heads of the annual *Briza maxima* are attractive in the border or cut for indoor decoration. It is planted here with the golden hop, *Humulus lupulus* 'Aureus'.

colour associations will then make an impression on our visual senses in one of three ways: they will either harmonize, contrast or clash. We are interested in the first two, and will probably want to avoid the third! It is actually quite difficult to concoct a clash, indeed it is virtually impossible to do so when using grasses. The danger is greater with stronger flower colours – pink with orange, for example.

The word harmony, normally implying the happy blending of any two or more factors, is rather more limited in discussing colour, signifying not just a combination of any two colours which look well together, but rather the association of closely related colours only. Using the colour circle (Fig. 2) you will observe the three primary colours and their three intermediaries. Harmony is the result of the juxtaposition

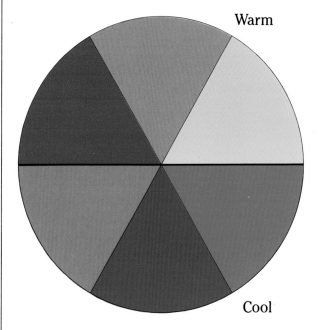

Fig. 2 The Colour Circle.

19

of any two adjoining colours. Contrasting colours are any that are not immediate neighbours to each other. (A further factor is introduced by the dark line which divides the warm colours from the cold.) Both harmonious and contrasting associations are satisfying to the eye. They differ in that harmonies are generally easier on the eye and therefore restful, whilst contrasts make a more forceful statement.

We also talked of balance – several restful shapes balanced by one vertical one. The principle of balance applies equally to the use of colour. The most satisfying planting is probably achieved by a basically harmonious arrangement, punctuated by a limited number of contrasts. These will bring that necessary sparkle to what could become a rather flat scheme, but should not be so numerous as to detract from the overall serenity of the panorama.

Yet another factor plays its part inasmuch as, in our painting with plants, colours vary not only in hue but also in lightness and darkness. We can have two colours which harmonize, but because one is light and the other dark we have a contrast of tone – say pale yellow with dark green. So our harmonies and contrasts may be of colour or tone, or both. It will be recognized that the possible permutations of tone, form, habit, height, leaf character, season of interest, limitations as to soil or situation etc. are boundless and there is no end to the scope for creativity, experiment, reassessment and rearrangement.

Grass borders for foliage effect

We will imagine that what you have read thus far has so enthused you that you have decided to set aside a border or part of the garden exclusively for grasses. Actually this represents a swing to the other extreme wherein we are severely limiting our palette. How-

ever, experimenting with a restricted palette can be an interesting exercise for an artist, and the same is true for the gardener. Besides, if we are especially interested in a group of plants why should we not devote an area to them? There are alpine fanatics, rose or chrysanthemum fanatics, heather and conifer fanatics, even (with due respect to them) vegetable fanatics – and their gardens reflect their passion. We grass enthusiasts (graminophiles?) need not be outdone. Yet we must recognize that we are presenting ourselves with a real challenge. We want our border to be, not just a collection of grasses, but a lively, interesting entity. Yes, we are back with contrast, and yet we have drastically limited the scope for that contrast. We have no deep purple, no silver, few really dark tones, and somewhat limited variety in leaf shape – certainly nothing rounded. We must therefore exploit to the fullest such contrast as is available. Select contrasting tones and colours, leaf and flower forms, and height, form and habit. Successful grass gardens can be achieved. The recent garden festivals at Stoke and Glasgow in the U.K. both featured examples, the one at the latter, in 1988, attracting particular comment.

Now, at last, let's get down to some real planting! We will assume that we have the luxury of an empty border in a position in good light with reasonable soil, already dug over and with some organic matter incorporated, just waiting for some plants.

The following scheme would suit a border of some 2 m (6½ ft) by 15 m (50 ft) and, in its entirety, would make a magnificent planting of grasses, illustrating the whole gamut of possible colour and shape variations. However, a border this size devoted to grasses would not be suited to every garden – or gardener – and so it is divided into three sections, each with its own harmonizing colour schemes (Figs. 3–5). Of course any of the associations could be used, right down to just two or three of the varieties suggested.

RED, YELLOW AND OLIVE HUES

Colourwise we will start with the yellows and yellow variegations, reds, olives, and greens which are on the yellow side rather than the blue. It is usually best to select the larger specimen-type plants first. (At this point I must mention that there are a number of grasses whose form dictates that they be used in this fashion – that is, as single specimens standing free from the nudging attention of any immediate neighbours which would only blur their individuality. Those which display a strongly symmetrical shape, hemispherical like *Helictotrichon sempervirens*, or erect and outward arching like *Carex buchananii*, can only be fully appreciated in isolation. Underplanting is quite permissible so long as it remains strictly round the feet of the specimen.) There is a choice of yellow variegated specimen grasses; we will employ two in this section of the border, and will aim for a contrast of form between them. *Phragmites australis* 'Variegatus' and *Miscanthus sinensis* 'Zebrinus' are both nicely erect plants, so one of those will suit us. If we choose the former we must be alert to its wandering habits, especially if the soil is at all damp, and either contain it or be prepared to remove growth that exceeds its allotted limits. We can next choose between two grasses of arching shape to contrast with the vertical nature of the miscanthus or phragmites. Either *Spartina pectinata* 'Aureamarginata' or *Cortaderia selloana* 'Gold Band', the golden variegated pampas grass, will be fine. The above caution with regard to the phragmites may be echoed in the case of the spartina, although it is more controllable, and in an especially cold region the pampas grass may need some winter protection, particularly while it is young. Both offer good flower heads as a bonus. Between these two we will use the shorter *Miscanthus sinensis purpurascens* whose leaves will become an increasingly strong reddish purple as the summer advances.

There are three fine yellow colours to come next, used in bold patches in front of and adjacent to these taller plants. Bowles' golden grass, *Milium effusum* 'Aureum', is virtually evergreen, and golden yellow in all its parts including the open flower panicles and the stems on which they are held. The bamboo, *Pleioblastus viridistriatus*, will attain 1 m (3 ft) or so, but not so much when the recommended spring cut-back is given to maintain the wonderful quality of the bright yellow-striped leaves. *Carex elata* 'Aurea' is Bowles' golden sedge, certainly the strongest yellow of all the grass-like plants. Between these last two is the tussock grass, *Chionochloa rubra*, a tuft of arching narrow foliage of a unique colour, a sort of orange ochre, grey-blue within the inrolled leaves, more brassy in winter. It is intriguing close up but nothing very special from a distance except when associated with bright yellow, when it really comes into its own. The rich red-browns of *Uncinia rubra* and *Uncinia uncinata* would look splendid in the foreground here, although they might not survive in colder areas without protection.

Carex kaloides and *Schoenus pauciflorus* are sedges of pale orange-brown and maroon respectively, both fairly erect, the former with flat leaves, the latter more needle-like. In absolute contrast to the schoenus is the beautiful, low-growing *Hakonechloa macra* 'Aureola' with soft lax leaves of yellow and green overlaying one another as the clump gently spreads. A reddish flush usually suffuses the leaves as autumn arrives. Between the two strong yellows of milium and pleioblastus is the dramatic rich red of the Japanese blood grass, *Imperata cylindrica* 'Rubra'. The amount of red increases as the summer progresses and, especially in the grouping here suggested, the effect is positively stunning. *Carex dipsacea* is a rich olive, almost bronzy in full sun, and with narrow arching leaves. It will stand out nicely in front of the yellow bamboo. It adjoins a front-of-border carpet of the

useful foxtail grass, *Alopecurus pratensis* 'Aureo-variegatus'. Clip this one back in early summer before it flowers and it will remain low, dense and bright. To the right of the sedge is space for a form of the wood-rush, *Luzula sylvatica* 'Hohe Tatra'. In all its varieties the woodrush is a valuable evergreen, holding its own in most situations including those which are often a problem, such as dry shade. This variety produces broad, rich green leaves during the summer which are pleasing enough, but it is in the winter that it really comes up trumps, transforming itself into a mound of strong, shiny yellow, a constantly glowing treat through the bleaker months of the year. *Carex testacea* forms arching tufts of a paler olive than *C. dipsacea*, with bright orange overtones in full sun, and looks good against yellow. Its neighbour is a fairly recent variety of the native wavy hair grass, *Deschampsia flexuosa* 'Tatra Gold', quite delightful in flower, although it is the lovely clean, brightest yellow-green of the narrow leaves that is the main attraction of this cultivar. Sadly, though, if you garden on shallow chalk, this is one of the few grasses that will not be happy with you. Against this I have chosen another fine variety of the woodrush, *Luzula sylvatica* 'Marginata', a smart plant with a narrow cream margin to the deep green leaves. The palm branch sedge, *Carex muskingumensis*, with bright, fresh green leaves completes this group. If your soil is on the damper side of average there are no plants included here that would complain – in fact several of them would be that much happier.

Harmonizing colours, contrasting leaf shapes: the sedges *Carex dipsacea* (left) and *C. morrowii* 'Fisher's Form' (right) with *Tolmiea menziesii* 'Taff's Gold' (centre) and *Lonicera nitida* 'Baggesen's Gold'.

BLUE, GREEN AND CREAM HUES

We proceed now to a more relaxed scheme and, leaving the yellow end of the spectrum we move to the mid- and blue-greens and to cream and white variegation which gives the added sparkle that is always welcome.

Two more varieties of the stately miscanthus family are the specimen plants in this section. *Miscanthus sinensis* 'Silver Feather' is the taller: erect and then arching, with drooping leaves, and with fine autumn inflorescences. *M.s.* 'Gracillimus' is less erect, and with very narrow grey-green, white mid-ribbed leaves, curled at the tips. An alternative in this position (except in colder areas) would be the white-variegated pampas grass, *Cortaderia selloana* 'Silver Stripe'. *Carex pendula* makes quite an impression with shiny rich green, arching leaves and long flowering culms with drooping heads. You may have seen it growing wild in damp woodland. Three markedly variegated forms appear next, all with foliage to about the 60 cm (2 ft) mark, and all of which are of spreading habit and will therefore need watching. (Additionally they will all grow in damp soil if required, the second and third even in water.) *Carex riparia* 'Variegata' can look almost white, with just a narrow green edge; *Phalaris arundinacea* 'Feesey's Form' is broader leaved and even whiter; and *Glyceria maxima* 'Variegata' is of essentially cream appearance, being only narrowly green-striped. The new growth is pleasingly pink-flushed. Each of these will create bright pools between the miscanthus specimens. The middle rank plants start with the snowy woodrush, *Luzula nivea*, so named on account of its white flower heads. The evergreen leaves are dark green and hairy-edged. *Carex oshimensis* 'Variegata' makes evergreen clumps of neatly green and white striped leaves. *Stipa gigantea* is a fairly low plant but sends up marvellous open plumes to 2 m (6 ft) or so. Broad leaves are the valuable

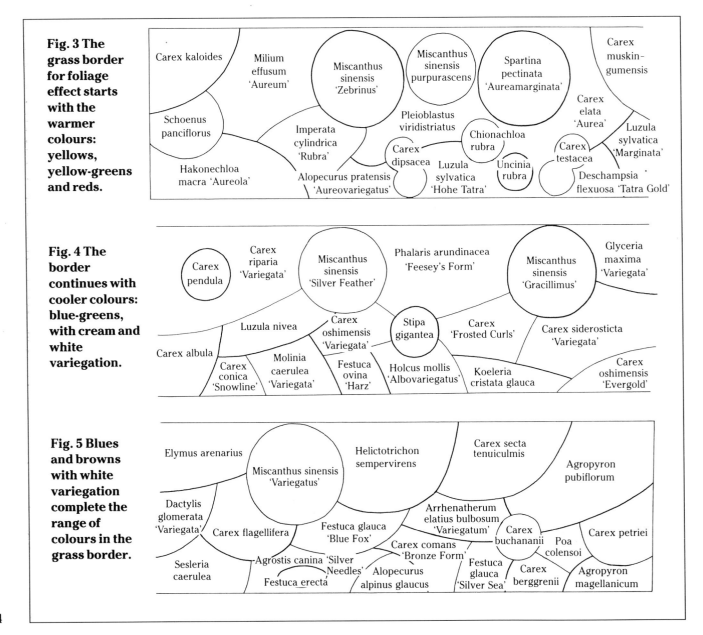

Fig. 3 The grass border for foliage effect starts with the warmer colours: yellows, yellow-greens and reds.

Fig. 4 The border continues with cooler colours: blue-greens, with cream and white variegation.

Fig. 5 Blues and browns with white variegation complete the range of colours in the grass border.

feature of the next sedge, *Carex siderosticta* 'Variegata'. They may exceed 2.5 m (1 in) in width, and have white stripes. In this plan they are deliberately used against the almost hair-like tresses of *Carex* 'Frosted Curls'. The colour in this variety is a silvery green. Other suggestions for this spot would be the pale grey-green *Sesleria nitida* or the blue-green quaking grass, *Briza media*. The colour of *Carex albula* is an unusual whitish green and its narrow but tough leaves form a symmetrical mop-head. *Carex conica* 'Snowline' is a charming little plant, neat, evergreen, and with a narrow white margin. It is used here at the front of the border, but it would also be at home in a rock garden, sink, or even in a pot on a well-lit window sill indoors.

Molinia caerulea 'Variegata' is a beauty. Fairly erect, it makes a tight, well-behaved clump with cream stripes to the blades. The flowers often show touches of purple and are held on cream stems. It will show up well placed in front of the darker green snowy woodrush. It will also contrast nicely with the blue-green, purple tipped tussocks of *Festuca glauca* 'Harz'. The next plant is the mat-forming *Holcus mollis* 'Albovariegatus' with soft, short leaves edged white. An alternative would be *Agrostis canina* 'Silver Needles' whose name aptly describes its narrowest of white-edged blades. *Koeleria cristata glauca* makes tidy blue-green mounds topped with pleasing flower heads. To round off this section we have another evergreen sedge, *Carex oshimensis* 'Evergold'. This is a particularly attractive variety with long narrow leaves arching out to form a low mound, and displaying a prominent creamy yellow median stripe. (This plant, too, will grow indoors in good light.)

BLUES AND BROWNS WITH WHITE VARIEGATION

The progression of colours moves on finally to the fascinating combination of blues and browns, with the additional crispness of white variegation. Assuming that this is a continuous border, the creams of the last section are matched with blues in this one – again a lovely association. *Miscanthus sinensis* 'Variegatus' is the specimen here, the best of the taller variegated grasses, surging upwards before cascading out. As an alternative on a slightly smaller scale we could select the new *Calamagrostis arundinacea* 'Overdam'. This has white-variegated leaves, usually pink-flushed, and delightful feathery flower heads, purplish becoming greyish pink and reaching 1.2 m (4 ft). *Elymus arenarius* is robust and vigorous and will need restraining, but it is a splendid grey-blue with stiff blue flower heads turning buff. The other taller specimen is *Carex buchananii*, with narrow red-brown foliage rising erectly to about 75 cm ($2\frac{1}{2}$ ft) and curling at the tips. Differing shades of brown are contributed by *Carex flagellifera*, a flat mid-brown, *Carex secta tenuiculmis*, slightly larger and a deep warm brown, the grey-brown *Carex comans* 'Bronze Form', and *Carex petriei*, erect and curly-topped like *Carex buchananii* but shorter and with wider blades of pinky brown. Quite different is the tiny *Carex berggrenii*, whose 5 cm (2 in) blunt leaves are a metallic grey-brown. Among these the blues are interspersed. *Helictotrichon sempervirens* is a fine steely blue forming a hemisphere of stiff leaves. The marvellous silvery blue of *Agropyron pubiflorum* is the most showy of all. It forms fairly erect clumps which spread only gently, retains its colour (though somewhat more green) through the winter, and seems just as good in shade. Coming down in height to about 25 cm (10 in) we have the narrow in-rolled leaves of *Poa colensoi*, which are of a good blue and again manage to maintain a good appearance right through winter. Of similar height and making a tight clump of pale blue is *Festuca glauca* 'Blue Fox', while *Festuca glauca* 'Silver Sea' is more compact and of a more silvery blue. The nature of *Festuca erecta* is given away by its name. This

Falkland Islander makes a low vertical accent of a subtle leaden blue, and in our scheme stands out of a carpet of *Agrostis canina* 'Silver Needles' mentioned earlier. *Alopecurus alpinus glaucus* features unusual colouring in that the silvery blue is often distinctly purplish. Two patches of cleanest white variegation are provided by *Dactylis glomerata* 'Variegata' and *Arrhenatherum elatius bulbosum* 'Variegatum'. The leaves of *Sesleria caerulea*, the blue moor grass, are held in a somewhat horizontal plane and form a very dense clump. The upper surface is blue, the undersides dark green, and a pleasing feature of the plant is its very early purple flower. Possibly the most brilliant pale blue leaves of any garden plant are displayed by *Agropyron magellanicum*. It is really startling and provides a fitting finale to our grass border.

Grass borders for flower effect

The wonderful range of foliage colour among the grasses and their lengthy period of interest mean that a border planted along the lines just discussed, including as it does a significant number of evergreens, will offer something to enjoy the year round, and the many differing flower heads will come and go as a bonus. An area given over primarily to floral effect is a different story, as any cultivator of the more traditional herbaceous border will know. The glory of summer may have as its price a decidedly uninspiring winter aspect. But what choice is there? The plants which sport those incredible extravaganzas of colour, so carefully selected and astutely juxtaposed for maximum summer impact, are rarely renowned for any great contribution outside that season. Thankfully the grasses are more obliging. For a start, as observed earlier, the period of interest of the inflorescences of the grasses is, more often than not, far longer, as flower heads mature to seed heads with little, if any, diminution of appeal. It is therefore no difficult task to achieve a progressive overlap of floral interest. Secondly the stems and foliage of many of the grasses fade most gracefully and may readily be left standing through the autumn and winter. The obligatory cutback of the herbaceous border to forestall the untidy mess that would inevitably follow instantly creates a yawning void where we have become accustomed to bulk and form. If we have made use of grasses those factors may be retained and enjoyed, along with the accompanying rustling sound effects. When it finally becomes necessary to cut them down in late winter the anticipation of spring and imminent regrowth is perfectly adequate to tide one over a brief period of bareness.

A current trend in North America is to use grasses in a way that recalls the heady days of the pioneering spirit, in creating 'prairie gardens'. In its simplest and therefore perhaps most evocative form this involves the mass planting of just one type of grass. The resulting sea of waving flower heads is a joy to behold. Extending the range to two or three, or at most four species adds variety to the scene. Now this idea assumes that a fair amount of space is available, and probably in an informal, even wild, setting, and most of us are not favoured with such suitable expanses. However, the principle of this concept is one that is wholly applicable to far more restricted settings. So, where it is the effect of the flower heads that is to be the main feature try this suggestion: use just two, three or four species, and fill the available space with these.

Here is one grouping (Fig. 6) that would provide interest from mid summer right into winter – and

The glorious colouring of *Carex elata* 'Aurea' is a spectacular feature in this yellow and green scheme.

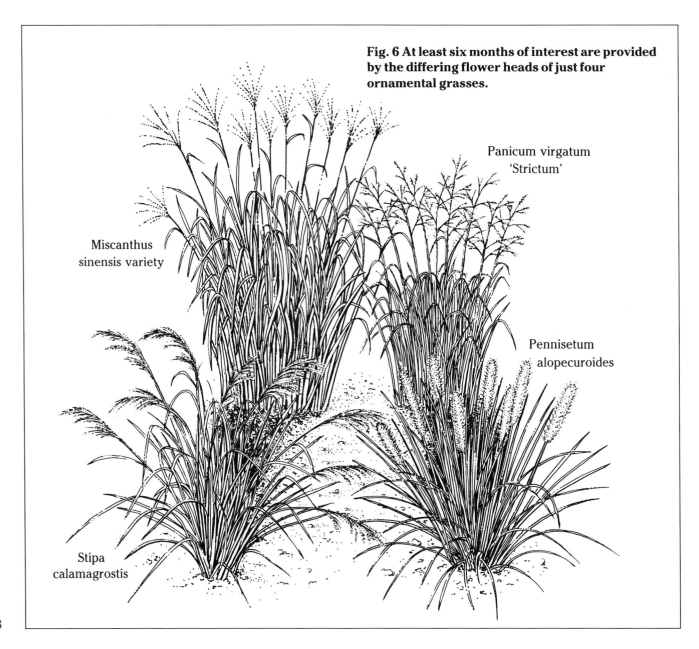

Fig. 6 At least six months of interest are provided by the differing flower heads of just four ornamental grasses.

Panicum virgatum 'Strictum'

Miscanthus sinensis variety

Pennisetum alopecuroides

Stipa calamagrostis

that's quite something. An area just 2 m (6½ ft) square would accommodate one plant of each of the four types suggested, two behind and two in front. At the back use any *Miscanthus sinensis* variety, except perhaps *M.s.* 'Variegatus', *purpurascens* and 'Gracillimus' which are not always reliably free-flowering, especially after a poor summer. Most varieties come into flower only in early autumn, although some, like 'Kleine Fontane' are much earlier. There are also variations in plant size ('Gracillimus' types are usually smaller), openness of the flower head (widest in 'Silberspinne') and flower colour – some, such as 'Kascade', 'Graziella', 'Malepartus' and 'Rotsilber' manifesting striking glistening red colouring, especially in the earlier stages of flowering.

Next to the finger-like miscanthus plumes, the airy, wide-open heads of *Panicum virgatum* 'Strictum' will create a pleasing contrast. They will appear at about the same time, starting tight but soon opening right out and, like miscanthus, will stand into the winter months. A fine autumn show of bright yellow leaves will add to the display, unless you select as an alternative one of the purplish-red varieties, whose leaves become progressively deeper in colour, and with chestnut brown flowers and seed heads.

Coming forward to the spot in front of the miscanthus we can find a home for one of the longest flowering of the grasses, *Stipa calamagrostis*. Once the flowers have made their initial appearance in early summer they just keep coming, right through until the autumn, and then stand well into winter. Furthermore they are a delight to behold – long, loose, feathery panicles, glinting with silver at first. This, and a pennisetum, our other foreground grass, are more arching in habit and will hide any tendency of the two varieties behind to look a little bare from the knees down. We could use *Pennisetum alopecuroides* in one of its selected forms, or *P. orientale*. These bear attractive bottle-brush or caterpillar flowers in profusion from late summer. They will, however, need some winter protection in colder areas.

An alternative combination (Fig. 7), with a similar long period of interest, could comprise varieties of *Calamagrostis* × *acutiflora*, *Molinia caerulea* and *Deschampsia caespitosa*. The calamagrostis is a grass of bolt upright habit and appears in the varieties 'Stricta' and 'Karl Foerster', with long, narrow, feathery inflorescences. This may be placed back left, flanked by a *Molinia caerulea arundinacea* form such as 'Karl Foerster' (a second reference to this worthy gentleman prompts a mention of his leading role in raising ornamental grasses and selecting many of the fine forms available to us today), or 'Windspeil', bearing graceful open flower heads.

The foreground feature in this group is *Deschampsia caespitosa*, the beautiful native tufted hair grass. The evergreen basal leaf clump is dwarfed from mid summer onwards by a hazy cloud of delicate flowers which progress through seed-bearing and remain into winter. Variations in flower colour are reflected in several varietal names, 'Bronzeschleier', 'Goldgehaenge' and others. If one plant were placed centrally in this arrangement room would be left either side which might well be occupied by the perennial quaking grass, *Briza media*, or its annual counterparts, *B. maxima* and *B. minor*, each with pendant trembling lockets in various sizes. Seeds of these latter two, scattered in spring will quickly produce flowering plants, as would those of the charming little hare's tail, *Lagurus ovatus*, with fluffy fat spikes.

There are a number of annual grasses which, like their popular bedding counterparts in a dazzling array of multifloriferous and grandifloriferous varieties, put in a really hard summer's work. They are easily raised from seed and should certainly not be forgotten as an alternative when there are spaces to be filled.

Other associations may be worked out following the basic idea of using flower heads which contrast as starkly as you wish. If more space is available try to resist the temptation to add a plethora of further types. The subtlety of these inflorescences allows for, perhaps even calls for, the multiple planting of each different type, and especially of the smaller ones. In restricted areas the same strategy can be adopted on a reduced scale using grasses of correspondingly smaller stature.

Mix-and-match planting

If we are to be perfectly realistic we must acknowledge that most of us are unlikely to devote large areas of our gardens solely to grasses, much as we might like them. In practice the suggested grass border for foliage effect, while demonstrating both the feasibility of such a concept and a logical way of progressing through the colour range, is more likely to be referred to for ideas on small-scale associations, perhaps just two, three or four grasses in a group. The point has been constantly reiterated that the best effect will be achieved by drawing from as many of the available plant types as possible. Without doubt the grasses are shown off to their best advantage in the company of the more traditional rounder-leaved plants. The converse must also be true. This would seem to lead us to the conclusion that we can plant our grasses with any and all other types of plant, virtually indiscriminately, and be assured of success. Believe it or not there is much truth in this. I have often made the comment in response to requests for suggestions on

The evergreen sedge *Carex flagellifera* is planted here with a hosta, nicely demonstrating that blue and brown are good companions.

livening up a part of the garden that 'isn't quite right', 'Plant a grass or two' and it invariably works wonders, at least from the point of view of form and texture.

However, by adopting a more considered approach we can expect even better results. For a start, there is one group of plants that will do no favours to grasses, nor receive any in return, and that is those with similarly linear or sword-like leaves – crocosmias, irises, kniphofias, hemerocallis and the like. Secondly, form and texture are only part of the story, and not usually the most important part. The earlier comment that colour is the factor that normally makes the first and greatest impression on our senses is worth repeating here. So colour, along with the closely related tone, will be the first consideration in the suggestions that follow, with form and texture as close runners-up.

Here are some colour-coordinated schemes in the mix-and-match style which draw on a variety of plant types, but in each of which a significant contribution is made by the grasses. Once again, just a small segment of one of the plans may provide inspiration for a gap that needs filling.

FIGURE 8: PURPLE, PINK, SILVER, BLUE AND WHITE-VARIEGATED

This scheme employs colours that are not well represented among the grasses, i.e. silver, deep purple and pink. Blue and white-variegated grasses are added in a mutually flattering association. There is barely a hint of green to be seen.

1. *Elaeagnus commutata.*
   Large silver shrub.
2. *Acer negundo* 'Flamingo'.
   Beautiful pink- and white-variegated large shrub or small tree.
3. *Cotinus coggygria* 'Royal Purple'.
   The best rich purple large shrub.

31

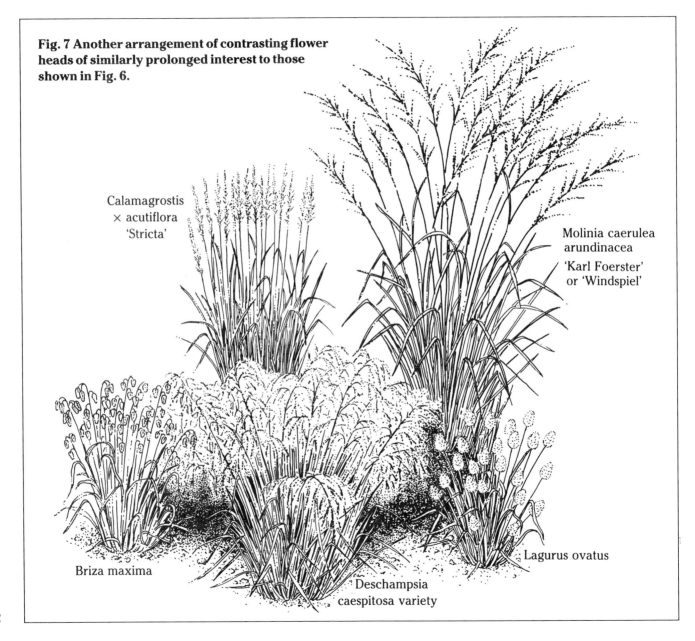

Fig. 7 Another arrangement of contrasting flower heads of similarly prolonged interest to those shown in Fig. 6.

Calamagrostis × acutiflora 'Stricta'

Molinia caerulea arundinacea 'Karl Foerster' or 'Windspiel'

Briza maxima

Deschampsia caespitosa variety

Lagurus ovatus

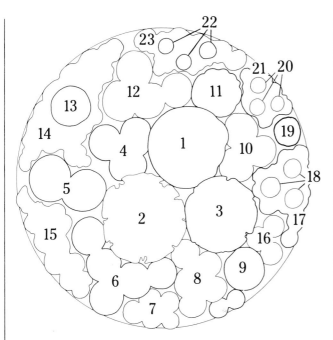

Fig. 8 Purple, pink, silver, blue and white-variegated.

4 *Elymus arenarius*.
Erect medium grey-blue grass.
5 *Euphorbia dulcis* 'Chameleon'.
Splendid new low euphorbia with deep purple leaves and flowers.
6 *Artemisia ludoviciana latiloba*.
Upright very silver perennial.
7 *Sedum* 'Vera Jameson'.
Greyish-purple round leaves. Rose red flowers.
8 *Agropyron pubiflorum*.
Erect silvery blue grass.
9 *Spiraea* × *vanhouttei* 'Pink Ice'.
New low shrub. Delicate pink new leaves, becoming cream then white-variegated. White flowers.

10 *Calamagrostis arundinacea* 'Overdam'.
New white-variegated, purple-flushed erect grass. Purplish, then greyish pink flower heads.
11 *Ruta graveolens* 'Jackman's Blue'.
Low evergreen mound of blue divided foliage.
12 *Phalaris arundinacea* 'Tricolor'.
White-variegated grass with good reddish purple flush.
13 *Helictotrichon sempervirens*.
Steely blue evergreen grass. Graceful flower heads.
14 *Stachys byzantina* 'Silver Carpet'.
Non-flowering woolly leaved silver carpeter.
15 *Agropyron magellanicum*.
Intense pale blue-leaved grass.
16 *Agrostis canina* 'Silver Needles'.
Prostrate white-variegated grass. Purplish flowers.
17 *Ajuga reptans* 'Burgundy Glow'.
Magenta, cream and grey-green carpeter. Blue flowers. Interplanted with *Alopecurus alpinus*. Low, silvery purple-blue grass.
18 *Arrhenatherum elatius bulbosum* 'Variegatum'.
Cleanly white-striped grass.
19 *Hebe* 'Wingletye' or *Hebe pinguifolia* 'Pagei'.
Low bluish evergreen shrubs.
20 *Festuca glauca* 'Blue Fox' or *Poa colensoi*.
Short blue grasses.
21 *Artemisia stelleriana prostrata*.
Silvery white carpeter.
22 *Festuca erecta*.
Stiffly straight-leaved blue-green grass.
23 *Holcus mollis* 'Albovariegatus'.
White-variegated carpeting grass.

FIGURE 9: MID- TO YELLOW-GREEN, OLIVE, YELLOW AND YELLOW-VARIEGATED

Always a fresh and cheerful mix, this scheme includes a preponderance of fairly intense colours. Cream

33

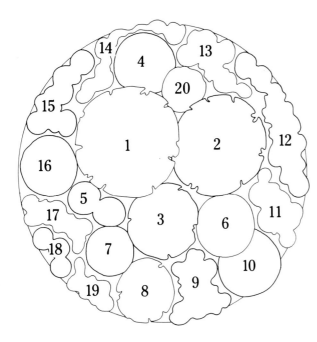

Fig. 9 Mid- to yellow-green, olive, yellow and yellow-variegated.

variegation may be added, being considerably softer. Some find the dramatic contrast of purple a welcome inclusion – for many it is just too extreme. Yellow, cream and white flowers would add further interest.

1. *Elaeagnus pungens* 'Maculata'.
   Yellow-variegated evergreen shrub.
2. *Sambucus racemosa* 'Plumosa Aurea' or 'Sutherland Gold'.
   Splendid yellow cut-leaved shrub.
3. *Lonicera nitida* 'Baggesen's Gold'.
   Evergreen shrub with tiny gold leaves.
4. *Spartina pectinata* 'Aureamarginata'.
   Yellow-margined arching grass.
5. *Lysimachia ciliata*.

Perennial with brown new leaves and pale yellow flowers.
6. *Miscanthus sinensis* 'Zebrinus'.
   Stately grass with yellow cross-banded leaves.
7. *Chionochloa rubra*.
   Evergreen arching grass with unusual brassy leaves.
8. *Choisya ternata* 'Sundance'.
   Bright yellow evergreen shrub.
9. *Carex testacea*.
   Orange-over-olive leaves. Evergreen sedge.
10. *Juniperus communis* 'Depressa Aurea'.
   Prostrate conifer. Yellow in summer, bronze in winter.
11. *Carex dipsacea*.
   Leaves bronze over deep olive. Evergreen sedge.
12. *Carex fortunei* 'Fisher's Form'.
   Evergreen sedge with cream-edged leaves.
13. *Carex muskingumensis*.
   Fresh green-leaved sedge.
14. *Carex elata* 'Aurea'.
   Bowles' golden sedge, very bright. Or *Milium effusum* 'Aureum', Bowles' golden grass.
15. *Alopecurus pratensis* 'Aureovariegatus', or *Hakonechloa macra* 'Aureola'.
   Bright yellow-variegated grasses.
16. *Deschampsia caespitosa* 'Goldgehaenge'.
   Grass with haze of golden flowers over green leaves.
17. *Lamium maculatum* 'Cannon's Gold'.
   Dead nettle with all-yellow leaves.
18. *Carex umbrosa* 'The Beatles'.
   Neat, low sedge.

Another evergreen sedge, *Carex oshimensis* 'Evergold' contrasts superbly with the broad, dark leaves of *Ajuga reptans* 'Braunherz'.

19 *Hedera helix* 'Angularis Aurea'.
Ivy with bright yellow young leaves.

20 *Foeniculum vulgare purpureum*.
Bronze fennel. Bronze-black leaves. Yellow flowers.

FIGURE 10: GREY- AND BLUE-GREEN, BLUE, SILVER, WHITE- AND CREAM-VARIEGATED

It is important to keep to the blue side of mid-green and exclude any hint of yellow. The inclusion of cream variegation may seem a contradiction, but it is actually a most effective addition. This is an essentially pale-toned scheme, which could be used in light shade if necessary.

Fig. 10 Grey- and blue-green, blue, silver, white and cream-variegated.

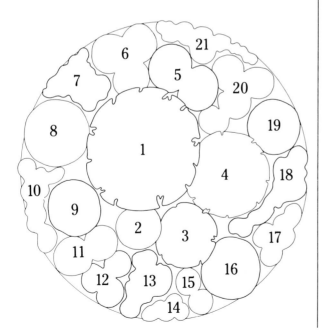

1. *Cornus alba* 'Elegantissima'.
White-variegated shrub.

2 *Juniperus scopulorum* 'Skyrocket'.
Tall narrow grey-green conifer.

3 *Miscanthus sinensis* 'Variegatus'.
Handsome grass with striking white variegation.

4 *Olearia macrodonta*.
New Zealand holly, evergreen olive-grey leaves.

5 *Hebe carnosula*.
Small blue evergreen leaves. White flowers.

6 *Tanacetum densum amani*.
Low evergreen mound of deeply cut silver leaves.

7 *Carex conica* 'Snowline'.
Neat low evergreen sedge. Deep green leaves with white margins.

8 *Dorycnium hirsutum*.
Soft grey-green leaves. Pinky white flowers.

9 *Stipa calamagrostis*.
Attractive free-flowering grass.

10 *Festuca glauca* 'Harz'.
Blue-green grass with purple tips.

11 *Phalaris arundinacea* 'Feesey's Form'.
Grass with strongly white-variegated leaves.

12 *Hosta* 'Halcyon'.
Good blue heart-shaped leaves.

13 *Molinia caerulea* 'Variegata'.
Neat cream-variegated grass.

14 *Carex atrata*.
Grey-green leaved sedge.

15 *Artemisia pontica*.
Perennial with soft feathery grey-green leaves.

16 *Cotoneaster horizontalis* 'Variegatus'.
Herringbone cotoneaster with small white-edged leaves.

17 *Carex oshimensis* 'Evergold'.
Evergreen sedge with broad central cream stripe.

18 *Agropyron pubiflorum*.
Erect silvery blue grass.

19 *Helichrysum serotinum.*
   Curry plant. Intensely silver low shrub. Cream buds. Remove yellow flowers as they open.
20 *Glyceria maxima* 'Variegata' or *Phalaris arundinacea* 'Luteopicta'.
   Cream-variegated grasses.
21 *Koeleria cristata glauca.*
   Blue-grey leaves. Nice grassy flower heads.

FIGURE 11: YELLOW, BROWN, ORANGE AND RED

A planting scheme exclusively of these colours gives a most vivid, almost fiery colour scheme, strengthened still further with bright yellow or red flowers if desired. Botanical purples may be added and some unusual effects may be achieved when these are used next to browns.

1 *Berberis thunbergii* 'Gold Ring'.
   Reddish-leaved shrub with narrow gold margin.
2 *Foeniculum vulgare purpureum.*
   Fennel with bronze-black leaves. Yellow flowers.
3 *Choisya ternata* 'Sundance' or *Ilex crenata* 'Golden Gem'.
   Evergreen shiny yellow-leaved shrubs.
4 *Thuya occidentalis* 'Rheingold'.
   Orange conifer.
5 *Milium effusum* 'Aureum'.
   Yellow grass.
6 *Carex flagellifera.*
   Brown mop-headed evergreen sedge.
7 *Carex elata* 'Aurea'.
   Bright yellow-leaved sedge.
8 *Carex buchananii.*
   Foxy-brown evergreen sedge.
9 *Juniperus communis* 'Depressa Aurea'.
   Prostrate conifer, yellow in summer, bronze in winter.
10 *Acorus gramineus* 'Ogon'.
   Evergreen fans of bright yellow grass-like leaves.

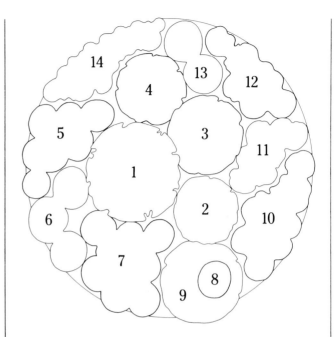

Fig. 11 Yellow, brown, orange and red.

11 *Lychnis × arkwrightii.*
   Very dark purplish leaves. Vermilion flowers.
12 *Alopecurus pratensis* 'Aureovariegatus' or *Hakonechloa macra* 'Aureola'.
   Bright yellow-variegated grasses.
13 *Carex secta tenuiculmis.*
   Deep warm brown evergreen sedge.
14 *Acaena inermis* 'Copper Carpet'.
   Bronze cut-leaved evergreen ground cover.

FIGURE 12: YELLOW, BLUE AND SILVER

This colour scheme is pretty if perhaps somewhat bland inasmuch as there is a virtual equality of light value. We are therefore deprived of the contrast of light and dark which add sparkle and crispness to the scene. However it is still very pleasing.

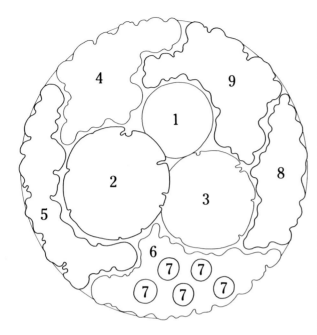

Fig. 12 Yellow, blue and silver.

1. *Chamaecyparis lawsoniana* 'Pembury Blue'.
   Good blue pyramidal conifer.
2. *Senecio* 'Sunshine'.
   Silver-grey evergreen shrub. Yellow flowers.
3. *Lonicera nitida* 'Baggesen's Gold'.
   Evergreen shrub with tiny yellow leaves.
4. *Carex elata* 'Aurea'.
   Bright yellow sedge.
5. *Agropyron magellanicum.*
   Intense pale blue-leaved grass.
6. *Artemisia stelleriana prostrata.*
   Silver-white carpeter.
7. *Festuca glauca* 'Blue Fox' or *Poa colensoi.*
   Short blue tufted grasses. Evergreen.
8. *Agropyron pubiflorum.*
   Erect silvery blue grass.

9 *Artemisia ludoviciana latiloba.*
  Upright very silver perennial.

FIGURE 13: BLUE, BROWN AND SILVER

Another wonderful combination, for which we are heavily dependent on grasses.

1 *Artemisia stelleriana prostrata.*
  Silver-white carpeter.
2 *Alopecurus alpinus glaucus.*
  Low, silvery purple-blue grass.
3 *Cerastium tomentosum columnae.*
  Compact form of the silver-leaved, white-flowering snow-in-summer.
4 *Carex buchananii.*
  Foxy-brown evergreen sedge.
5 *Hebe pinguifolia* 'Pagei'.
  Low blue-grey evergreen shrub. White flowers.
6 *Agropyron magellanicum.*
  Intense pale blue-leaved grass.
7 *Chamaecyparis lawsoniana* 'Blue Surprise' or 'Van Pelt's Blue'.
  Grey-blue conifers with a hint of purple in winter.
8 *Santolina chamaecyparissus.*
  Cotton lavender. Evergreen silver leaves.
9 *Carex comans* 'Bronze Form'.
  Fine-leaved, mop-headed, evergreen brown sedge.
10 *Festuca glauca* 'Blue Fox'.
   Short blue tufted grass. Evergreen.
11 *Acaena inermis* 'Copper Carpet'.
   Bronze cut-leaved ground cover.
12 *Carex berggrenii.*
   Dwarf grey-brown sedge.

The broad leaved *Carex siderosticta* 'Variegata' is the centrepiece of this semi-shady planting. It also looks good in a pot.

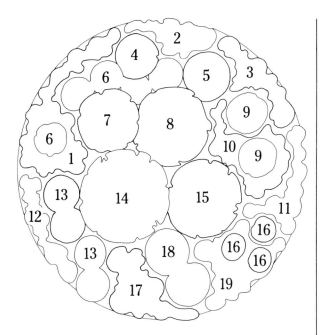

Fig. 13 Blue, brown and silver.

13 *Agropyron pubiflorum.*
Erect silver-blue grass.
14 *Weigela florida* 'Foliis Purpureis'.
Unusual brown-leaved shrub, dusky pink flowers.
15 *Ruta graveolens* 'Jackman's Blue'.
Low evergreen mound of blue divided foliage.
16 *Carex petriei.*
Small, curly-topped, pinky-brown evergreen sedge.
17 *Stachys byzantina* 'Silver Carpet'.
Non-flowering woolly-leaved silver carpeter.
18 *Artemisia ludoviciana latiloba.*
Upright very silver perennial.
19 *Festuca glauca* 'Silver Sea'.
Small tufted silver-blue grass.

'Grasses and ground cover' borders

This idea is a compromise between the all-grass and 'mix-and-match' suggestions. It gives the pre-eminence to grasses which many enthusiasts may desire, but still retains the facility for using colours, shapes and textures which are complementary and additional to those of the grasses. My somewhat arbitrary rule is that with grasses up to about 60 cm (2 ft) in height the surrounding ground cover plants should not exceed 10 cm (4 in); with grasses between 60 cm (2 ft) and 1.2 m (4 ft) the maximum ground cover height should be around 20–25 cm (8–10in); and nothing in excess of 45 cm (18 in) should be planted with grasses over 1.2 m (4 ft). Some of these useful plants, many evergreen, are listed below.

GROUND COVER PLANTS UP TO 10 cm (4 in):

Acaena, in variety
Achillea, in variety
Ajuga, in variety
Antennaria dioica
Artemisia stelleriana prostrata
Cerastium tomentosum columnae
Cotula potentilloides
Fragaria × *ananassa* 'Variegata'
Glechoma hederacea 'Variegata'
Hedera, in variety
Lamium, in variety
Lysimachia nummularia 'Aurea'
Parthenocissus henryana
Sagina subulata 'Aurea'
Sedum, in variety
Thymus, in variety
Trifolium repens 'Quadrifolium Purpurascens'
Vinca minor varieties
Waldsteinia ternata

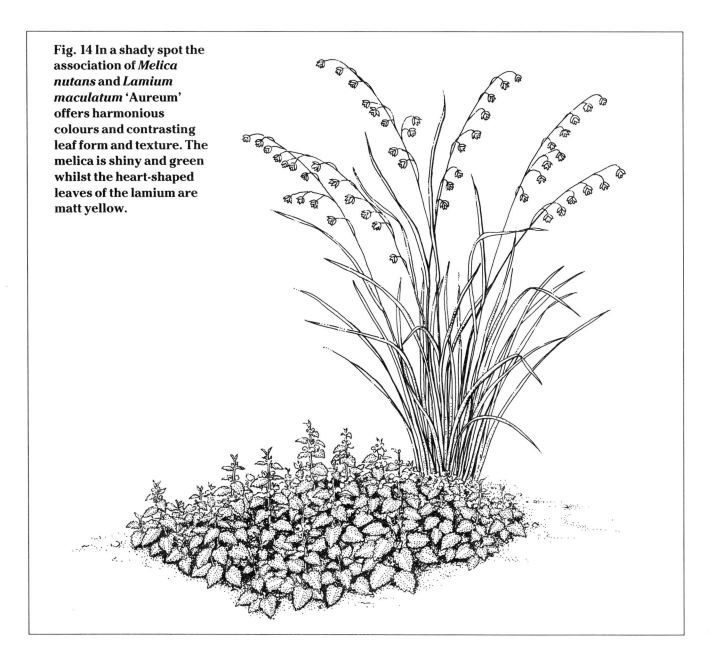

Fig. 14 In a shady spot the association of *Melica nutans* and *Lamium maculatum* 'Aureum' offers harmonious colours and contrasting leaf form and texture. The melica is shiny and green whilst the heart-shaped leaves of the lamium are matt yellow.

GROUND COVER PLANTS 10–25 cm (4–10 in):

Aegopodium podagraria 'Variegata'
Bergenia, in variety
Euphorbia myrsinites
Hebe, in variety
Hosta, in variety
Houttuynia cordata 'Chameleon'
Juniperus, in variety
Ophiopogon planiscapus 'Nigrescens'
Pachysandra terminalis
Ranunculus repens 'Joe's Golden'
Sedum, in variety
Stachys byzantinum 'Silver Carpet'
Symphytum 'Goldsmith'
Tolmeia menziesii 'Taff's Gold'
Vinca major varieties

GROUND COVER PLANTS 25–45 cm (10–18 in):

Artemisia, in variety
Cotoneaster horizontalis 'Variegatus'
Euonymus fortunei varieties
Euphorbia dulcis 'Chameleon'
Hebe, in variety
Heuchera micrantha 'Palace Purple'
Rubus idaeus 'Aureus'

Shingle, rock and paved gardens

We have so far discussed only the interrelationship of plant material and have discovered numerous poss-ibilities for pleasing associations. When we embark on a consideration of the materials with which we surround our plants we enter a whole new realm of contrast: that between those living, moving plants, ever-changing through the transient phases of growth and development, and inanimate materials, natural or manufactured, solid, consistent, unyielding. We can cover the surface with shingle or pebbles. We can assemble arrangements of rocks, larger stones or boulders. We may introduce either stepping stones or a continuous path through the planting, with a further choice of natural stone or the straight lines, rectangles and precise circles of man-made slabs. Round wooden stepping stones and log-roll edging are also available.

Furthermore, many of these items may be obtained in a choice of colours and tones, offering further poss-ibilities of harmony and/or contrast with each other and with our grasses. For example, the warm creamy buffs of Cotswold and similar stone combine delight-fully with the blue and brown grasses. Given a piece of virgin ground and a free choice as to how to use it I would go for a 'grasses and ground cover' planting in a rock, shingle and natural paving setting.

It is important to remember when gardening in this style that the additional features are there to be seen as part of the overall scheme, so we are not looking for total ground coverage as we might be ordinarily. Unplanted areas should be allowed for with the plant-ing arranged in carefully orchestrated, probably irregular, groups and masses.

For a virtually maintenance-free garden lay medium- to heavy-gauge polythene over the soil and pierce holes in it for planting. Then spread shingle over the entire surface. You will need to bear in mind that creeping plants will not be able to spread (should you want them to) unless you progressively cut away the polythene from around them.

A close-up of the glorious plumes of *Cortaderia selloana*.

42

GRASSES FOR HOT, SUNNY SPOTS

Whether or not the 'greenhouse effect' will give us hotter, drier summers remains to be seen. However, some soils are very light and free-draining, and just a few days of sunshine will leave them parched. The situation can be improved by incorporating plenty of organic material into the soil at planting time, and by mulching, but here are some grasses that will tolerate such conditions, once established:

Agropyron magellanicum
Alopecurus alpinus glaucus
Alopecurus lanatus
Corynephorus canescens
Elymus arenarius
Eragrostis curvula
Festuca – all blue-leaved types
Festuca mairei
Helictotrichon sempervirens
Koeleria cristata glauca
Lagurus ovatus
Pennisetum species
Poa colensoi
Poa labillardieri
Poa buchananii

GRASSES FOR SHADE AND WOODLAND

There is a grass for every conceivable situation, and the woodlands and glades of the temperate regions of the world are well-endowed in this respect. Although most of them will grow quite satisfactorily in more open conditions, and most of those whose natural habitat is open and sunny are not in the least troubled by light shade, it is good to know that there are a number of grasses that we can draw from when it comes to rather darker situations. Grasses happy in shade include:

Agrostis canina 'Silver Needles'
Aira elegantissima
Bromus ramosus
Carex conica 'Snowline'
Carex fraserii
Carex grayi
Carex morrowii varieties
Carex oshimensis varieties
Carex pendula
Carex plantaginea
Carex saxatilis 'Variegata'
Carex siderosticta 'Variegata'
Carex umbrosa 'The Beatles'
Chasmanthium latifolium
Chusquea couleou
Deschampsia caespitosa and varieties
Deschampsia flexuosa
Holcus mollis 'Albovariegatus'
Luzula species and varieties
Melica nutans
Melica uniflora 'Variegata'
Milium effusum 'Aureum'
Pleioblastus – several species
Poa chaixii
Sasa veitchii

GRASSES FOR WET SOILS, BOG GARDENS AND WATER

The tolerance that some of the grasses show to a diverse range of conditions is no better demonstrated than by some of those mentioned here, which will grow in ordinary, even rather dry soils but also with their feet in water. The following plants will all tolerate wet soil, and those marked with an asterisk will grow in water:

Arundo donax*
Carex elata 'Aurea'*
Carex grayi*

Fig. 15 *Achnatherum brachytrichum* features attractive purplish-grey flower heads on erect culms. The small leaves of *Cotoneaster horizontalis* 'Variegatus' are an interesting feature around the base of the grass.

Carex riparia 'Variegata'
Carex trifida
Cyperus species*
*Eriophorum angustifolium**
Glyceria maxima 'Variegata'*
Juncus effusus and varieties*
Molinia caerulea and varieties
Phalaris arundinacea varieties*
Phragmites australis and varieties*
Scirpus lacustris 'Albescens'*
Scirpus lacustris tabernaemontani 'Zebrinus'*
Spartina pectinata 'Aureamarginata'
Typha species and varieties*

GRASSES IN POTS AND TUBS

Certain grasses are well suited to growing this way, notably those of arching growth or of hemispherical shape which manifest a degree of symmetry. The classic grass for pot culture is undoubtedly *Hakonechloa macra* 'Aureola'. *Carex siderosticta* 'Variegata' is also good, as are the taller *Carex buchananii* and *Chionochloa rubra*. The evergreen dome of *Helictotrichon sempervirens* is another candidate. Note any particular cultural requirements of the plants concerned, and water assiduously.

Coloured lawns?

If common grasses can be sown, or laid as turf, and maintained as a lawn the question must be put: What about a coloured or variegated lawn? Well, it is theoretically possible. Probably the best ornamental grass

Cortaderia selloana 'Gold Band' in the grass border at the Royal Horticultural Society's garden at Wisley, Surrey. The blue Elymus arenarius is behind.

for this purpose would be *Agrostis canina* 'Silver Needles' which would give an interesting whitish effect and could be mown as normal once well established. Other candidates which could be cut with the mower blades set high would be *Holcus mollis* 'Albovariegatus' and *Alopecurus pratensis* 'Aureovariegatus'. Other small grasses such as the blue festucas could be close-planted and trimmed over with shears, but would not take too kindly to being walked on regularly. There are drawbacks, however. Every seed that lands on your 'lawn' is going to develop into a green-leaved grass or weed – and will stand out like a sore thumb, unless you are prepared for some diligent hand-weeding. May I suggest that this fascinating idea be implemented in small areas only, that can easily be given the attention they will need, not as true lawns for walking on, but perhaps to create pleasing geometrical patterns using several varieties adjoining each other. Alternatively squares, rectangles, T- or L-shapes etc. could be formed within areas of paving slabs and filled with a single type of grass.

Grasses for indoor arrangements

It should not be forgotten that the grasses are a major source of valuable material for cutting for fresh or dried arrangements and many respond well to dyeing. Books on the subject will give details relevant to particular species, but as a general rule flower heads need to be cut well before maturity to lessen the risk of shattering.

CHAPTER 4

Cultivation and Propagation

Choosing your grasses

Preliminary to cultivation and propagation is the purchase of plants. There is considerable confusion over the correct naming of some of the grasses and, particularly, the sedges. While it is good to see more grasses appearing in garden centres I am appalled at the lack of attention to this important matter. If you want to be fairly sure of getting the plants you particularly want it is far safer to go to a specialist nursery.

Whatever the source, if you are buying on the spot rather than ordering by post (which will often be the only way of acquiring many less common varieties) be sure that the plant is well established in its pot. It is the responsibility of the nurseryman to care for the plant until that point is reached, and it will then have the best chance of growing away happily.

Tidying up

The grasses as a group are no more difficult to cultivate than any other group of plants. Some species have particular needs and these should be noted when purchasing and planting. Spring and summer planting especially should be followed by conscientious attention to watering until the plant is established. Subsequent care more or less parallels that for most herbaceous perennials. If they are evergreen the leaves may suffer some winter damage, particularly at the tips. Such damage should be removed in the spring, but do not trim away more than is necessary for a tidy appearance. Through the course of the year a certain amount of dead foliage may become apparent, and often this can be removed by running the fingers, or even a comb, through the plant. In other cases it may be necessary to take the scissors to them and simply cut away what is obviously dead material.

With herbaceous grasses you have a choice: if you like the look of the withered foliage, stems and flower heads, leave them standing until late winter and then cut them to the ground. Don't delay any longer or you may damage the emerging leaves which will then have flat rather than pointed tips! If they are untidy in the autumn, and will clearly do nothing to enhance the winter garden, cut them right back straight away. One or two varieties become untidy around the middle of the summer or lose the brightness of their variegation and, as with some other perennial plants, these can be cut back for a second flush of attractive growth. Some forms of *Phalaris arundinacea* are examples of this.

Coping with invasive species

It is worth giving a little thought to those grasses which are invasive spreaders, inasmuch as several of them are most desirable plants which one may be loathe to exclude from the garden. Apart from this latter course it seems to me that one has three options.

First, they can be grown in pots or tubs. They are ideal candidates for this type of culture because they will fill out their containers quite rapidly. Watering must never be neglected, and feeding during the growing season will help. If necessary they can be split and started again every two or three years.

A second possibility is to surround the plant with material that will physically restrict its spread. A plastic bucket with the bottom cut out is often suggested, and will be fine for small to medium grasses. A plastic dustbin treated the same way and cut in half would give two reasonably sized barriers. Again it will be beneficial to lift the plant and divide and replant it at intervals as it may die out in the middle. Larger and stronger grasses are best restrained with concrete, to a depth of at least 45 cm ($1\frac{1}{2}$ ft). This could be cast *in situ*, or paving slabs could be arranged vertically in the ground, or one might be able to obtain sections of large-diameter concrete pipework. None of these are particularly straightforward options, and in appropriate circumstances mowing round clumps of such grasses will be equally effective.

The third choice is a simple one: acknowledge the threat when you first put the plant in, allot it a realistic area into which to spread, and then take your spade to it regularly, cutting it back to well within this area. In practice, one inspection when new growth has appeared in spring and another in mid to late summer will be the most that is required. I know of no grass that will put up much resistance to a sharp spade at such regular intervals. This controlled approach effectively eliminates any potential threat. After all, when you think about it, it is the established, out of control clump which always causes the real headache.

Pests and diseases

Pests and diseases are rare. Occasional greenfly or blackfly attacks may be dealt with using any of the appropriate products available. Just two or three grasses may be seen to be developing rust on the leaves in summer, especially if you live in an area where cereal crops are grown. Early treatment with Dithane 945 (produced by PBI), following the manufacturer's recommendations closely, should solve the problem and prevent spread.

Propagating grasses

Propagation by division is possible with all the grasses and, apart from a tiny minority where cuttings are an alternative, is the only method in the case of the many cultivars. So far as the species are concerned seed will give larger numbers without disturbing the plant itself. The collecting of seed may be carried out when the flower head turns brown, first checking that the individual seeds come away readily. This may usually be done either by pulling at the hairlike awns or by gently rubbing the head between fingers and thumb where awns are absent. Store the seed in an envelope in a cool, dry place until required.

Seeds of hardy annuals may be sown in spring direct into prepared soil where they are to grow. Autumn sowing of many of them will produce more robust plants the following year, often flowering earlier. Half-hardy annuals and perennials should be sown and pricked out in warmth in early spring, hardened off and planted out after frosts, or sown direct into the ground in late spring. Hardy perennials can be sown in pots or trays in spring, preferably with some protection, pricked out and planted in their growing positions when large enough.

Division is simply a matter of lifting the grass and prising the crown apart. The more robust grasses may call for the use of secateurs, a spade or the traditional two forks back to back. Usually several small plants may be obtained by this method. The divisions should be replanted and watered regularly until established. The safest time to carry out division is from when the plants start into growth in spring until mid summer. Winter division will usually result in dead plants.

49

CHAPTER 5

Alphabetical List of Grasses, Bamboos, Rushes and Sedges

The following list gives basic details of a significant proportion of the grasses that are available to gardeners. As well as the botanical name and plant family the dimensions of each plant are given. '*L*' indicates the height and then spread of the leaves, '*Fl*' the flowering height. To ascertain the planting distance between two plants add together the figures given for their spread and divide by two. Unless otherwise stated it may be assumed that the plant is perennial and herbaceous, i.e. dying back in winter.

ACHNATHERUM

Achnatherum brachytrichum   Gramineae
L 60 × 60 cm (2 × 2 ft)   *Fl* 1.2 m (4 ft)
Green leaves, slightly bronze-flushed when young, emerge early, forming a loose drooping clump. Long, narrow, slightly open feathery heads of greyish pink adorn the plant in late summer and autumn. A striking feature in the middle of the border between shrubs. Most soils in sun or light shade.

AGROPYRON

Agropyron magellanicum   Gramineae
L 15 × 30 cm (6 × 12 in)   *Fl* 15 cm (6 in)
Leaves, culms and flower spikes all boast the brightest pale blue colour, and are all decidedly lax. Use for foreground planting, perhaps as a carpet around the erect reddish brown *Carex buchananii* and backed by a silver shrub. Evergreen, but less blue in winter. Average to dry soil in sun.

Agropyron pubiflorum   Gramineae
L 50 × 30 cm (20 × 12 in)   *Fl* 75 cm (2½ ft)
An erect, slowly expanding clump of lovely pale silvery blue, looking good with brown sedges, with purple-leaved plants such as the new *Euphorbia dulcis* 'Chameleon', or with pink-flushed variegated plants like *Spiraea × vanhouttei* 'Pink Ice'. Add silver for a finishing touch. Inflorescence in early to mid summer of no great significance. Leaves greener in winter. Most soils which are not too heavy, in sun or moderate shade.

AGROSTIS

***Agrostis canina* 'Silver Needles'**   Gramineae
L 8 × 30 cm (3 × 12 in)   *Fl* 20 cm (8 in)
Discovered as a chance seedling in a London garden by John Fielding who brought me a plant, and I suggested the name 'Silver Needles' as appropriately describing its narrowest of blades, each cleanly white-margined. A delightful little grass, essentially horizontal in habit, spreading gently by overground runners, forming a carpet, though never a nuisance. Use to surround erect deeper green grasses. Good in small pots, also for variegated lawns, planted 20 cm (8 in) apart. Flowers are a delight, airy clouds of

It would be difficult to devise a more dramatic contrast than that between *Glyceria maxima* 'Variegata' and *Ligularia dentata* 'Othello' – both best in damp soil.

glistening purple – but best removed before seeding. Most soils (except the driest) in sun or moderate shade.

Agrostis nebulosa   Gramineae
L 20 × 30 cm (8 × 12 in)   *Fl* 35 cm (14 in)

Cloud bent grass. Annual. Good for drying, attractive in the border. Well justifies its common name, producing veritable clouds of small spikelets in delicate open heads floating over smaller clumps of foliage. Sow seeds *in situ* in spring or autumn to flower in mid to late summer. Average to good soil in sun.

AIRA

Aira elegantissima   Gramineae
L 15 × 20 cm (6 × 8 in)   *Fl* 30 cm (12 in)

Hair grass. Another annual, named after its very fine leaves. If anything more dainty than *Agrostis nebulosa* (above), with tiny silvery spikelets in loose, airy panicles in early summer. A popular grass for drying. Sow seeds where they are to grow in ordinary to light soil in sun or part shade.

ALOPECURUS

Alopecurus alpinus glaucus   Gramineae
L 8 × 15 cm (3 × 6 in)   *Fl* 20 cm (8 in)

Blue foxtail grass. I find myself increasingly fond of this little grass. Silvery overtones to subtle grey-blue to purple leaves. Gentle spreader, not a nuisance, forming very loose clumps – a few leaves here, a tuft there – hence ideal for intermingling with other carpeters, such as the magenta and cream *Ajuga reptans* 'Burgundy Glow', or any of the prostrate silvers. Stubby purplish flower spikes in spring. Average to light soil in sun.

Alopecurus lanatus   Gramineae
L 8 × 12 cm (3 × 5 in)   *Fl* 15 cm (6 in)

Silver foxtail grass. A little gem with thickly woolly leaves appearing silvery over grey. Hardy but dislikes winter wet so some care is required. Alpine house cultivation would be ideal; otherwise protect with the traditional overhead pane of glass, or simply surround with a good layer of grit in a quick-draining soil in full sun. Suitable for sinks. Small fat flower heads in spring.

Alopecurus pratensis 'Aureovariegatus'
Gramineae
L 40 × 30 cm (16 × 12 in)   *Fl* 70 cm (28 in)

Golden foxtail grass. Slowly spreading clump of marvellous yellow leaves, narrowly striped green, forming a bright patch at the front of the border. Trim once or twice in summer to keep it low (if desired). Cylindrical flower spikes, early summer. Average to good soil in sun or light shade. Use as a carpet round brown or olive sedges, or with any yellow-green or yellow-variegated shrub. The ground-hugging *Acaena inermis* 'Copper Carpet' is another splendid companion.

ANDROPOGON

Andropogon scoparius   Gramineae
L 90 × 45 cm (3 × 1½ ft)   *Fl* 1 m (3¼ ft)

Little blue stem. A grass of the North American plains, grown for its progressively changing foliage colour. Emerging in spring a pale grey green, it develops through purple in late summer to splendid foxy red tones in autumn. Erect growing, tight clump, topped in autumn by wispy inflorescences. Apparently hardy, but dislikes waterlogging in winter, and should therefore be planted in good but well-drained soil in sun.

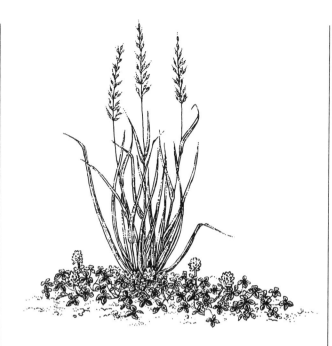

Fig. 16 The white-striped *Arrhenatherum elatius bulbosum* 'Variegatum' stands out of a carpet of the deep blackish-purple foliage of *Trifolium repens* 'Quadrifolium Purpurascens'.

ARRHENATHERUM

***Arrhenatherum elatius bulbosum* 'Variegatum'**
Gramineae
L 30 × 30 cm (1 × 1 ft)   *Fl* 45 cm (18 in)
Variegated bulbous oat grass, or onion couch. Possesses the strange characteristic of forming small bulbs at the base of the stems. Pure white-striped and margined grey-green leaves, very clean-looking. Narrow panicles in summer. Cut back for a second flush when foliage deteriorates after mid summer. Good in groups, or as specimens in a carpet of darker tones. Avoid the driest soils. Sun or part shade.

ARUNDO

Arundo donax Gramineae
L 4 × 1.5 m (13 × 5 ft)   *Fl* 4.5 m (15 ft)
Giant reed. From southern Europe, so requires a warm spot, in good to moist soil. Long grey-green leaves splay out from the stems, alternating in decidedly regular fashion. Use at the back of a large border or as a specimen, preferably cutting the stems right down each winter. Large late autumn flower heads not usually produced in cooler climates. Good examples may be seen in Britain at the Royal Horticultural Society's gardens at Wisley, Surrey.
***A.d* 'Variegata'** Surely one of the most striking of all the grasses, with broad creamy white margins and stripes. It is also decidedly tender, requiring virtually frost-free conditions. Try planting it in a pool with several inches of water above the crown. A shorter plant than the all-green species, to about 1.8 m (6 ft).
Arundo pliniana is of similar height, with short stiff leaves coming quickly to a point, giving a rather spiky appearance.

AVENA

Avena sterilis Gramineae
L 60 × 60 cm (2 × 2 ft)   *Fl* 1 m (3¼ ft)
Animated oat. Popular annual for the garden or indoor decoration, fresh, dried or dyed. Loose heads of nodding bristled spikelets, green becoming papery. Sow seeds *in situ* in spring or autumn. Any soil in sun.

BOTHRIOCHLOA

Bothriochloa caucasica Gramineae
L 40 × 60 cm (16 × 24 in)   *Fl* 75 cm (2½ ft)
Beard grass. Features reddish purple inflorescences in late summer, with arching foliage also turning reddish towards the autumn. Flower heads comprise multiple

54

Hakonechloa macra **'Aureola' is a choice grass for good soil,
here seen in a classic association with the broad leaves
of a blue hosta.**

A telling contrast of colour, tone and form, with *Holcus mollis* 'Albovariegatus' and the purplish, ribbed leaves of *Plantago major* 'Rubrifolia'.

spikes arising from the stem in quite erect fashion like a small miscanthus. Average soil in a sunny, sheltered spot. Protect in extreme cold, or save the seed and sow *in situ* in spring.

BOUTELOUA

Bouteloua gracilis   Gramineae
L 25 × 30 cm (10 × 12 in)   *Fl* 50 cm (20 in)
Mosquito grass or signal-arm grass. Appropriately descriptive common names refer to the flowering head, the rachis (main stalk) of which is held almost horizontally to the upright culm, and from which hang the densely packed brownish-purple spikelets and their dangling florets – a distinctive and most intriguing arrangement. Flowers through the summer. Dense clumps of narrow green leaves. Grasses of the open prairies of the Americas, requiring well-drained soil in sun.
Bouteloua curtipendula is larger and with erectly held inflorescences.

BRIZA

The botanical names of the three quaking grasses to consider progress not only alphabetically but in descending order of the size of the dangling spikelets, of value in the garden as they tremble and dance in the wind, and when cut as valuable contributors to dried flower arrangements.

Briza maxima   Gramineae
L 40 × 30 cm (16 × 12 in)   *Fl* 60 cm (2 ft)
The largest of the *Briza* family is an annual with light green, rather oblong spikelets in late spring and through the summer.

Briza media   Gramineae
L 40 × 30 cm (16 × 12 in)   *Fl* 60 cm (2 ft)
Perennial species. Heart-shaped purplish spikelets,

early to late summer, above a dense clump of somewhat blue-green leaves. Good soil in sun or part shade.

Briza minor   Gramineae
L 30 × 25 cm (12 × 10 in)   *Fl* 45 cm (1½ ft)
Lesser quaking grass. Smaller in all its parts but bearing a greater number of spikelets over a long period.

BROMUS

Bromus brizaeformis   Gramineae
L 50 × 30 cm (20 × 12 in)   *Fl* 75 cm (2½ ft)
Large heads of gracefully drooping spikelets similar to those of *Briza maxima* but longer and more pointed. May be picked and dried. Annual. Sow spring or autumn in any soil in sun or part shade, to flower throughout the summer.

Bromus macrostachys var. lanuginosus
Gramineae
L 25 × 20 cm (10 × 8 in)   *Fl* 50 cm (20 in)
The effect is almost woolly – quite different from the other species described here – with whitish heads of compact spikelets arising from a tight clump. Annual. Sow spring or autumn where it is to flower. Sun or part shade.

Bromus madritensis   Gramineae
L 30 × 25 cm (12 × 10 in)   *Fl* 60 cm (2 ft)
Compact brome. Long bristly spikes with strong reddish purple colouring in early summer. Annual. Sow spring or autumn in ordinary soil in sun or part shade.

Bromus ramosus   Gramineae
L 45 × 30 cm (1½ × 1 ft)   *Fl* 1 m (3¼ ft)
Wood or hairy brome. Perennial British native for more shady areas or, given good soil, in sun. Graceful in flower with open branched heads of drooping spikelets, narrow, bristled, and of rather wispy effect, mid and late summer.

CALAMOGROSTIS

Calamagrostis × acutiflora 'Karl Foerster'
Gramineae
L 100 × 45 cm (3¼ × 1½ ft)   *Fl* 1.5 m (5 ft)
Strongly vertical form, culms thrusting skywards topped by long, narrow, but slightly open, pinkish brown flower heads which remain well into the winter. Most soils, sun to part shade. Clump forming.

Calamagrostis arundinacea 'Overdam'
Gramineae
L 60 × 45 cm (2 × 1½ ft)   *Fl* 1.2 m (4 ft)
A most exciting recent introduction, this variegated grass is invaluable for the middle of the border. Leaves emerge with yellow margins and stripes, but before long, as they arch out from the rising culms, the yellow gives way to white with a pink flush, overlaid with a silky sheen. Flower heads in mid to late summer are purplish becoming greyish pink. Loose clump, gently spreading but not invasive. Average soil in sun or part shade.

CAREX

Carex albula   Cyperaceae
L 20 × 40 cm (8 × 16 in)   *Fl* 20 cm (8 in)
The first of some 35 members of this major family of sedges that we shall be considering is typical of several, mainly from New Zealand, which form a dense tuft of narrow evergreen leaves, starting erect, but soon lengthening and arching outwards in symmetrical fashion to produce a mop-head, or 'pudding-basin' hairstyle effect. I notice when exhibiting these species that there seems to be an irresistible urge to touch them and run one's fingers through their long tresses. They all need planting with a little space around them so that their form is not lost. Most are in distinctive colours, *C. albula* being on the smaller side and with unusual whitish green colouring, excellent in lighter-toned plantings. Insignificant flower spikes, of similar colour, produced mostly within the foliage. Easy in most soils in sun or light shade. Could be confused with *C. comans* or *C.* 'Frosted Curls'.

Carex atrata   Cyperaceae
L 15 × 30 cm (6 × 12 in)   *Fl* 30 cm (12 in)
Jet sedge. Quite an unusual colour for this family, rather glaucous blue-green. Leaves fairly broad, keeled, and tapering to a point. Rhizomatous, forming a slowly spreading clump, for the front of the border. Flower spikes oval, almost black. Any soil in sun or part shade.

Carex berggrenii   Cyperaceae
L 5 × 15 cm (2 × 6 in)   *Fl* 5 cm (2 in)
Interesting dwarf sedge, forming loose, slowly spreading clumps. Leaves a most unusual grey- to reddish brown with a metallic sheen, blunt-ended. Small brown flower spikes in mid summer. Good to moist soil in sun. Don't allow it to be swamped by neighbouring plants. There are also forms with grey-green leaves, both of normal width and much narrower.

Carex buchananii   Cyperaceae
L 75 × 60 cm (2½ × 2 ft)   *Fl* 75 cm (2½ ft)
One of several brown-leaved species whose names have become dreadfully confused in the trade. Actually quite distinctive, essentially erect form, arching outwards to produce a vase shape. Taller than the others with decidedly curled leaf tips. Narrow but extremely tough leaves, rich reddish brown, often with orange flecks. Brown flowering spikes through the summer on stems which may exceed 1 m (3¼ ft), curving back down to the ground. Easily grown in most soils in sun. Usually hardy but will require protection in extreme cold. Ideal for use as a vertical accent with lower-growing yellow-leaved plants.

Carex comans Cyperaceae
L 45 × 90 cm (1½ × 3 ft) Fl 45 cm (1½ ft)
A mop-headed sedge, larger than *C. albula* with very narrow pale green leaves. Mid summer flower spikes mostly hidden within the foliage. Most soils in sun. Evergreen.
C.c. 'Bronze Form' Of identical size and habit but pale warm brown in colour. Unusually effective with silver-leaved plants and the blue grasses.

Carex conica 'Snowline' Cyperaceae
L 12 × 15 cm (5 × 6 in) Fl 15 cm (6 in)
Neat and pleasing, if not especially showy, forming a low arching tuft. Narrow white margins to deep green leaves. Small flower spikes in early summer. Good soil in sun or shade. Protect in extreme cold. Evergreen.

Carex dipsacea Cyperaceae
L 45 × 45 cm (1½ × 1½ ft) Fl 45 cm (1½ ft)
An arching, clump-forming sedge of quite unusual colouring. Evergreen leaves deep olive green, overlaid with a bronzy glaze. Small black flower spikes in summer. Good soil in sun or shade, but note that the bronze colouring will be lost in insufficient light. Use in association with yellow and yellow-variegated plants.

Carex elata 'Aurea' Cyperaceae
L 70 × 45 cm (2¼ × 1½ ft) Fl 75 cm (2½ ft)
Bowles' golden sedge. Strongest yellow colouring of all the grass-like plants, with only narrow green margins and presenting a truly magnificent spectacle. Mr Bowles must have whooped with delight when he first came across this plant in the Norfolk Broads! Upper, male spikes brown, lower ones green, like fat caterpillars. Needs good to wet soil in sun. Would form a startling partnership with the Japanese blood grass, *Imperata cylindrica* 'Rubra'.

Carex firma 'Variegata' Cyperaceae
L 8 × 10 cm (3 × 4 in) Fl 10 cm (4 in)

A real dwarf, ideal for a sink or trough, or for a place in the rock garden free from the interference of neighbouring plants. Short, very stiff leaves quickly narrow to a point, shiny dark green with bold and clearly defined creamy yellow margins. Small dark brown flower spikes. Grow in limy soil, incorporating both grit and organic matter, in full sun.

Carex flagellifera Cyperaceae
L 45 × 90 cm (1½ × 3 ft) Fl 45 cm (1½ ft)
Another brown species which is frequently misnamed. Similar habit to *C. comans* 'Bronze Form', tufted, with arching leaves forming a mop-head, but the leaves broader, and of a warm, much redder brown, almost gingery. Avoid the driest soils, otherwise easy in a sunny position. Plant either with silvers and yellows or silvers and blues.

Carex fraseri Cyperaceae
L 50 × 45 cm (20 × 18 in) Fl 45 cm (1½ ft)
A distinctive sedge requiring moisture-retentive lime-free soil in shade. Small, round, bright white flowers appear in spring as the broad leaves unfurl.

Carex 'Frosted Curls' Cyperaceae
L 60 × 45 cm (2 × 1½ ft) Fl 60 cm (2 ft)
Similar in style to *C. albula* and *C. comans*, with narrow, pale green leaves giving a shiny, silvery effect, but much more erect, with curling tips. Pale green flower spikes inconspicuous within the foliage. Most soils in sun or part shade.

Carex grayi Cyperaceae
L 60 × 45 cm (2 × 1½ ft) Fl 60 cm (2 ft)
Mace sedge. Fascinating pale green seed heads,

An association of strong colours: the grass is *Imperata cylindrica* 'Rubra', with *Houttuynia cordata* 'Chameleon' and a yellow carpet of *Lysimachia nummularia* 'Aurea'.

Fig. 17 Interesting plant associations are possible in the garden pool. Here *Carex grayi* with its curious seed heads contrasts with the broad leaves and spathes of *Caltha palustris*, the bog arum.

knobbly and globe-shaped, a novel addition to cut flower arrangements. These develop through the summer above fairly broad, erect clumps of fresh green leaves which retain their colour well into the autumn. Grow in good soil in semi-shady border, or in water, or anything in between.

Carex kaloides   Cyperaceae
L 50 × 20 cm (20 × 8 in)   *Fl* 50 cm (20 in)
Shiny leaves vary between ochre and pale orange-

brown, fairly stiff and held upright, best displayed with yellows, interesting with blues. Pale brown inflorescences thin and insignificant. Good soil in sun.

***Carex morrowii* 'Fisher's Form'**   Cyperaceae
L 40 × 30 cm (16 × 12 in)   *Fl* 45 cm (1½ ft)
Particularly showy evergreen sedge. Broad, stiff leaves conspicuously margined and striped with cream. Flowering spikes pale green, appearing in spring. Good soil in sun or shade.
***C.m* 'Variegata'**   The greeny white stripes not very prominent but the plant as a whole makes a good feature with evergreen leaves of a particularly rich green. Flowers and growing preferences as above.

Carex muskingumensis   Cyperaceae
L 75 × 45 cm (2½ × 1½ ft)   *Fl* 75 cm (2½ ft)
Palm branch sedge. Cheerfully fresh green leaves differing from those of most of the sedges in that they arise from the erect culms rather than from the crown of the plant, the topmost group resembling palm branches. Forms gradually spreading clumps. Golden brown spikes in early summer of no great interest. Try using this carex with golden-leaved shrubs and perennials, especially hostas, in good soil in sun or part shade. The variety **'Wachtposten'** has yellow-green leaves.

***Carex ornithopoda* 'Variegata'**   Cyperaceae
L 15 × 25 cm (6 × 10 in)   *Fl* 20 cm (8 in)
A neat low tuft of green leaves with a central white stripe, for the front of the border, or the rock garden, in most soils in sun or part shade.

***Carex oshimensis* 'Evergold'**   Cyperaceae
L 25 × 38 cm (10 × 15 in)   *Fl* 30 cm (1 ft)
Deservedly one of the best known sedges. A low mound of long, arching, evergreen leaves, deep green with a broad central stripe of creamy yellow. In full sun this latter colour predominates, whereas in shade

the contrast between this and the green is more evident. Brown flowering spikes through spring. Good soil in sun or moderate shade, or even indoors as a pot plant in good light.

C.o. 'Variegata' has a white central stripe.

Carex pendula   Cyperaceae
L 60 × 90 cm (2 × 3 ft)   *Fl* 1.2 m (4 ft)
Drooping sedge. Terminal flowering spikes hang from long arching stems in early summer. Broad, keeled leaves, shiny green, blue underneath. Needs some space around it. Best in good soil in informal areas in part shade.

Carex petriei   Cyperaceae
L 25 × 15 cm (10 × 6 in)   *Fl* 25 cm (10 in)
Like a small *C. buchananii*, with narrow, erect leaves, curled at the tips, but the colour is different – pale pinky brown. A good short vertical accent plant amongst prostrate silvers and blues. Most soils in sun.

***Carex pilulifera* 'Tinney's Princess'**   Cyperaceae
L 10 × 15 cm (4 × 6 in)   *Fl* 15 cm (6 in)
A delightful form of a small British native, the pill sedge. Similar effect to that of the better known *C. oshimensis* 'Evergold', but in a far more delicate mode, with essentially creamy yellow leaves narrowly green-margined. Small flowers held above the neat mound of foliage in late spring and early summer. Any lime-free soil in sun or shade. Small enough for a sink or trough, or delightful at the border front or in the rock garden.

Carex plantaginea   Cyperaceae
L 15 × 30 cm (6 × 12 in)   *Fl* 20 cm (8 in)
Plantain-leaved sedge. A North American native. Broad, somewhat ribbed leaves up to 2.5 cm (1 in) in width, flushed reddish at the base. Yellow-brown flower spikes in spring. A plant for good soil in cooler situations in part or full shade.

***Carex riparia* 'Variegata'**   Cyperaceae
L 45 × 30 cm (1½ × 1 ft)   *Fl* 60 cm (2 ft)
One of the few sedges that could be called invasive, especially in moist conditions, but a most striking plant, especially in spring and early summer, with long, narrow leaves of pure white usually only narrowly margined with green. Flowering spikes, too, are quite smart, blackish, in late spring and early summer. Good to moist soil in sun or part shade.

***Carex saxatilis* 'Ski Run'**   Cyperaceae
L 10 × 15 cm (4 × 6 in)   *Fl* 10 cm (4 in)
An interesting small sedge with somewhat contorted leaves striped with white, and gradually forming a low carpet. Good to damp soil in some shade.

Carex secta   Cyperaceae
L 45 × 45 cm (1½ × 1½ ft)   *Fl* 45 cm (1½ ft)
An arching clump of narrow, evergreen leaves, splendid bright green, excellent with yellows. Tiny spikelets alternate up the flowering culms in early to mid summer. Good to moist soil in sun.

C.s. tenuiculmis   A recently available form with glowing, deep warm brown foliage.

***Carex siderosticta* 'Variegata'**   Cyperaceae
L 25 × 40 cm (10 × 16 in)   *Fl* 30 cm (1 ft)
A handsome, slowly spreading clump of broad, overlapping leaves, up to 2.5 cm (1 in) in width, white-margined and striped, and with a reddish flush at the base. Narrow flowering spikes in spring. Requires good soil in sun or, preferably, part shade.

Carex testacea   Cyperaceae
L 50 × 60 cm (20 × 24 in)   *Fl* 50 cm (20 in)
An eyecatching evergreen sedge from New Zealand. Might be confused with *C. dipsacea* but has narrower, longer leaves, paler olive, the upper parts which receive full light being orange. Flowering spikes differ, too, in being pale brown rather than black. Excellent

Lamarckia aurea **is an annual grass with intriguing downswept flower spikes.**

63

The evergreen woodrush *Luzula sylvatica* **'Marginata'**
softens the lines of the brick wall.

in company with yellows. Good to moist soil – in sun for the best colour.

Carex trifida   Cyperaceae
L 90×75 cm (3×2½ ft)   *Fl* 1.2 m (4 ft)
A robust sedge forming an erect clump of broad, grey-green evergreen leaves. The flowering culms hold multiple large brown spikes above the foliage from early to mid summer. A handsome plant for good soil through to boggy conditions.

Carex umbrosa 'The Beatles'   Cyperaceae
L 10×20 cm (4×8 in)   *Fl* 15 cm (6 in)
Imagine the heads of John, Paul, George and Ringo with hair dyed green and you have a fair impression of this small sedge! Gently spreading clump of narrow leaves topped by brown spikelets in spring. Happy in most soils in sun or shade, this neat plant is ideal for underplanting shrubs.

Carex uncifolia   Cyperaceae
L 8×12 in (3×5 in)   *Fl* 5 cm (2 in)
Dwarf sedge with narrow, curved leaves forming a dense tuft. Plants raised from seed may vary in their unusual colouring between reddish bronze and a paler pinkish bronze. Small flowering spikes hidden within the foliage. Good soil in a sink, rockery or at the front of the border with small blue grasses and creeping silver-leaved plants.

CHASMANTHIUM

Chasmanthium latifolium   Gramineae
L 100×45 cm (3¼×1½ ft)   *Fl* 1.2 m (4 ft)
Spangle grass. Interesting features are the 2 cm (¾ in) broad leaves (hence *latifolium*) splaying out from the culms, and the large ironed-flat spikelets which hang gracefully in open panicles. Full sun, so long as the soil is not too poor, through to moderate shade.

CHIONOCHLOA

Chionochloa conspicua   Gramineae
L 1×1 m (3¼×3¼ ft)   *Fl* 1.8 m (6 ft)
Chionochloas are the tussock grasses from New Zealand, all featuring dense clumps of arching foliage. This and the next species are the most ornamental

Fig. 18 The subtle colouring of *Chionochloa rubra* is best displayed against bright yellow, here represented in the foliage of the evergreen shrub *Choisya ternata* 'Sundance'. *Hedera helix* 'Angularis Aurea' creeps around its feet.

overall, reminiscent of the pampas grass but with more open, lax feathery plumes of creamy white, flowering early, from mid summer onwards. Good soil in sun, some shelter. Protect in extreme cold. Excellent dried.

Chionochloa flavescens   Gramineae
L 75 × 90 cm (2½ × 3 ft)   *Fl* 1.5 m (5 ft)
Similarly ornamental, generally smaller than *C. conspicua*, leaves unusual brownish green. The large, pale, airy panicles appear even earlier, from late spring in a good year, and remain right through until the autumn. Fine addition to dried arrangements. Similar cultural requirements to previous species.

Chionochloa rubra   Gramineae
L 75 × 60 cm (2½ × 2 ft)   *Fl* 75 cm (2½ ft)
Superficially drab, but closer inspection reveals subtle colouring: ochre upper surfaces to the narrow ever-green leaves, grey-blue undersides, and brassy tips. With its outward-arching form it requires a specimen position, and should be planted with bright yellows. Sparse mid summer flowers borne just above the clump. Average to moist soils in sun.

CHUSQUEA
Chusquea couleou   Gramineae (Bambuseae)
L 3.6 × 1.5 m (12 × 15 ft)
A tall Chilean bamboo with thick, solid culms of greenish yellow. The short leaves on densely clustered branches are an attractive feature, especially at the point when these are combined with the current year's shoots ready to burst from their protective white sheaths. These taller, clump-forming bamboos may be used as foreground or lawn specimens, much as a large shrub or small tree might be, or among lower shrubs at the back of a border. Good soil in sun or shade.

COIX
Coix lacryma-jobi   Gramineae
L 75 × 45 cm (2½ × 1½ ft)   *Fl* 75 cm (2½ ft)
Job's tears. Annual. The strange seed cases, hard, oval, pearly grey, may be removed when mature, dried and strung as beads. Leaves broad with a notably pale central vein. Treat as a half-hardy annual, sowing seed under glass in early spring and planting out after frosts in good soil in sun.

CORTADERIA
Cortaderia fulvida   Gramineae
L 1.2 × 2 m (4 × 6½ ft)   *Fl* 2 m (6½ ft)
Our look at the genus which includes the pampas grass starts with two earlier flowering relatives from New Zealand. This is the least hardy, and if grown away from milder areas should be protected through really cold spells. Evergreen foliage makes a tidy, tight clump. Flower heads appear from early summer, not nearly so full or shaggy as a pampas, but no less pleasing, and hanging slightly to one side. Colour pinkish cream, becoming ivory. Free-draining soil in sun.

Cortaderia richardii   Gramineae
L 1.5 × 1.8 m (5 × 6 ft)   *Fl* 2.5 m (8 ft)
Toe-toe. On a larger scale, with flower heads more sub-stantial than those of *C. fulvida*, whiter in colour, and carried on arching culms which reach way beyond the limits of the foliage beneath, requiring a space some 4–5 m (13–16 ft) in diameter. Well-drained soil in sun.

Cortaderia selloana   Gramineae
L 1.5 × 2 m (5 × 6 ft)   *Fl* 3 m (10 ft)
The familiar pampas grass, much used, sometimes unwisely so, in small suburban gardens, without due consideration to its ultimate size. Large, dense tussock of sharp-edged leaves topped by shaggy plumes in late summer and autumn, requiring considerable space in

65

66

Melica uniflora with its graceful beadlike flowers is planted
here with a fern in a shady setting which both prefer.

Yellow and blue are cheerful companions. *Milium effusum* **'Aureum' grows here in partial shade with self-sown forget-me-nots.**

which to be appreciated, but unrivalled when such space is available. Choose named varieties as seed-raised plants may bear disappointing flower heads. Dried heads are popular, but do pick them young for the best results. Well-drained soil in an open position. A cut-back in spring every two or three years will maintain a tidy appearance. Protect newly planted specimens from extreme cold in their first winter.

C.s. 'Gold Band' Smaller, with 1.5 m (5 ft) plumes over 1 × 1.5 m (3¼ × 5 ft) foliage. Increasingly effective as the summer progresses, with broadly yellow-margined leaves becoming steadily richer in colour. A splendid feature in the autumn garden.

C.s. 'Monstrosa' The largest variety, with enormous open, shaggy plumes to 2.7 m (9 ft) just above the 2 m (6½ ft) foliage clump.

C.s. 'Pumila' The best green-leaved pampas grass for the smaller garden. Flower heads more erect, almost cigar-shaped, and produced in abundance, 1.8 m (6 ft) over 1 m (3¼ ft) foliage.

C.s. 'Rendatleri' Carries large plumes of purplish pink to 3 m (10 ft).

C.s. 'Rosea' Features a distinct pink flush to the silvery plumes. Overall a little smaller, to 2.5 m (8 ft).

C.s. 'Silver Stripe' A white-margined cultivar which, like 'Gold Band', gets better as the weeks go by. Not quite so hardy as most, nor offering such an impressive floral display, but still an exciting feature plant. 1.8 m (6 ft) flowers over 1 m (3¾ ft) foliage.

C.s. 'Sunningdale Silver' Among the larger forms, this variety has no rivals when it comes to the quality of its 2.7 m (9 ft) flower heads, which are a joy to contemplate. Open, feathery and glistening creamy white, borne on sturdy, erect culms over 1.5 m (5 ft) foliage.

CORYNEPHORUS

Corynephorus canescens Gramineae
L 10 × 15 cm (4 × 6 in)   *Fl* 25 cm (10 in)

Grey hair grass. A plant of poor, thin soils. Dense tufts of stiff, narrow leaves, blue-grey with a hint of purple and reddish at the base. Small heads of reddish spikelets with unusual club-shaped bristles in mid summer. Sun.

CYPERUS

Cyperus eragrostis Cyperaceae
L 60 × 30 cm (2 × 1 ft)   *Fl* 75 cm (2½ ft)
From South America and related to the papyrus of warmer climes. Normally hardy. May be grown in the pond or in the border in good to wet soil but may require protection if extreme cold threatens. All parts of the plant are a fresh, bright green, including the attractive flowering spikelets borne in umbels from mid summer to early autumn, and which may be used to good effect in floral decoration. Easily raised from seed.

Cyperus longus Cyperaceae
L 100 × 60 cm (3¼ × 2 ft)   *Fl* 1 m (3¼ ft)
Galingale. Long, corrugated leaves of the shiniest mid-green. Pendulous, branched flowering heads carried on arching stems, bearing reddish brown spikelets, an asset to late summer flower arrangements. A rapid spreader in proportion to the dampness of the environment, and will grow in water. Will need containing in confined quarters.

DACTYLIS

***Dactylis glomerata* 'Variegata'** Gramineae
L 45 × 45 cm (1½ × 1½ ft)   *Fl* 75 cm (2½ ft)
Showy white-striped form of the British native cocks-foot. Dense arching clumps. One-sided panicles divided into groups of densely clustered green spikelets. Remove before seeds drop. Ordinary soil in sun or part shade.

DESCHAMPSIA

Deschampsia caespitosa Gramineae
L 50 × 50 cm (20 × 20 in)　*Fl* 1.2 m (4 ft)
Tufted hair grass. Undoubtedly one of the most graceful wild grasses. Clouds of hazy flower heads on erect
stems through the summer months in a variety of tints
and varying heights as reflected in the selected forms
which follow. These stand well into the winter and are
fine subjects for picking and drying. Leaves in a dense
evergreen clump. Tolerant of a range of conditions
from ordinary to damp soil, and sun to part-shade. Use
in woodland or natural gardens, or in the border for
contrast standing above lower plants or between
shrubs.
D.c. 'Bronzeschleier' ('Bronze Veil') The bronze
veil forms as the silvery green flowers mature.
D.c. 'Goldgehaenge' ('Golden Showers') Golden
yellow flowers appear a little later on 1 m (3¼ ft) stems.
D.c. 'Goldschleier' ('Golden Veil') Bright golden
yellow flower heads rise to about 1.2 m (4 ft).
D.c. 'Goldtau' ('Golden Dew') A more compact
form, with flower heads to about 75 cm (2½ ft) over
40 cm (16 in) foliage, with later flowers maturing to
warm gold.
D.c. vivipara Sometimes offered with the variety
name 'Fairy's Joke' attached. Something of an oddity,
the joke being that as the 'flowers' develop you realize
that they aren't flowers at all, but tiny plantlets which,
as they steadily enlarge, gradually weigh the culms
down until, from an erect 75 cm (2½ ft) start, they touch
the ground, sometimes rooting at the point of contact.
Such a form is technically described as viviparous.

Deschampsia flexuosa Gramineae
L 15 × 20 cm (6 × 8 in)　*Fl* 30 cm (1 ft)
Wavy hair grass. Also grows wild in Britain, in similar
conditions to its larger relative, and is tolerant of even
more shade. Prefers acid soils, so not suitable for

chalk. A smaller grass, leaves very narrow, bright mid-
green. The flowers are a graceful delight, purplish and
glistening with silver as they react to the gentlest
breeze, narrow at first, subsequently opening out. For
the front of the border, or in woodland settings.
D.f. 'Tatra Gold' A lovely variety with absolutely
the freshest and cleanest bright green yellow leaves
imaginable. A little less yellow as the season advances,
or if grown in too shady a spot.

ELYMUS

Elymus arenarius Gramineae
L 60 × 60 cm (2 × 2 ft)　*Fl* 1.2 m (4 ft)
Invasive, especially in dry soils, but none the less a fine
grass. Grey-blue leaves and stiff stems bearing long
wheat-like heads in summer. Effective with purples
and silvers, for example *Cotinus coggygria* 'Royal
Purple' and *Senecio* 'Sunshine', or with pink-flowered
grasses such as *Pennisetum orientale* and *P. villosum*.
Average to light soil in sun.

ERAGROSTIS

Eragrostis curvula Gramineae
L 75 × 60 cm (2½ × 2 ft)　*Fl* 90 cm (3 ft)
African, or weeping love grass. Needs a warm sunny
spot and a specimen position with space around it so
that the arching form of both the narrow foliage and
the long, open panicles of tiny spikelets, oddly
coloured dark greyish, may be fully appreciated.
Flowering period, late summer to early autumn.
Protect from extreme cold.

ERIOPHORUM

Eriophorum angustifolium Cyperaceae
L 30 × 30 cm (1 × 1 ft)　*Fl* 45 cm (1½ ft)
Cotton grass. Grown for the cotton wool effect of the
mass of soft hairs attached to the multiple seed heads.

Fig. 19 *Elymus arenarius* can be invasive, but its pale blue leaves and thrusting flower heads make it a splendid companion to larger shrubs, such as *Acer negundo* 'Flamingo' with its pink, white and green foliage. *Stachys lanata* 'Silver Carpet' provides appropriate ground cover.

Flowers late spring to early summer. Normally used as a marginal plant in ornamental pools, but will also grow in damp soil – but watch its creeping roots.

Eriophorum vaginatum Gramineae
L 30 × 15 cm (12 × 6 in)   *Fl* 45 cm (1½ ft)
Hare's tail. Flower heads slightly smaller than those of

E. angustifolium, and borne singly, in mid to late spring. Clump forming rather than running. Ordinary to damp soil will suffice.

FESTUCA

Festuca amethystina Gramineae
L 25 × 25 cm (10 × 10 in)   *Fl* 45 cm (1½ ft)
F. glauca and its varieties, discussed shortly, are the best known fescues or blue grasses. This species is similar but a little larger with pale grey-blue foliage, and flowering heads opening to a branched panicle in late spring and early summer. Well-drained soil in full sun.

Festuca erecta Gramineae
L 30 × 20 cm (12 × 8 in)   *Fl* 40 cm (16 in)
A grass of upright habit from the Falkland Islands. Narrow leaves rise stiffly like needles from a tight clump. The subtle leaden blue looks good standing out of a silver or white-variegated carpet. Late spring to early summer flowers are a typical, gradually opening small panicle. Average to dry soils in full sun.

Festuca eskia Gramineae
L 8 × 15 cm (3 × 6 in)   *Fl* 15 cm (6 in)
A soft little grass forming rich green, gently spreading mounds. More carpet-like if clipped back at regular intervals. Flowers, early to mid summer, green, turning brown. Average soil in sun. Fun with the cheerful hummocks of *Thymus* 'Archer's Gold'.

Festuca filiformis Gramineae
L 30 × 30 cm (1 ft × 1 ft)   *Fl* 30 cm (1 ft)
Somewhat similar to *F. eskia* but on a decidedly larger scale.

***Miscanthus sinensis* 'Silver Feather' features in a pleasant autumn scene.**

Festuca glacialis Gramineae
L 5 × 10 cm (2 × 4 in) *Fl* 13 cm (5 in)
A tiny, dense mound of very narrow whitish blue-green leaves. Violet spikelets, mid to late summer. Well-drained soil in full sun.

***Festuca glauca* 'Azurit'** Gramineae
L 20 × 15 cm (8 × 6 in) *Fl* 35 cm (14 in)
All of the *F. glauca* varieties are useful for edging or as foreground groups, either massed or spaced as small specimens, interplanted with prostrate silvers. Best foliage colour is maintained by dividing the clumps every two to three years. Small, open flower heads appear in early summer. 'Azurit' is taller than most, and an excellent blue.

***F.g.* 'Blaufuchs' ('Blue Fox')** and ***F.g.* 'Blauglut' ('Blue Glow')** Both good blue selections, 15 cm (6 in) high.
***F.g.* coxii** Another good blue, but rarely flowers, if you should find that an advantage.
***F.g.* 'Harz'** This gives an altogether different effect, with blue-green leaves tipped purple and greyish purple flower heads 15 cm (6 in) high.
***F.g.* 'Meerblau' ('Ocean Blue')** Features blue-green foliage, to a similar height to *F.g.* 'Harz'.
***F.g.* 'Minima'** There is some doubt as to whether this tiny grass is a festuca at all. Wiry little leaves, 5 × 10 cm (2 × 4 in), of very pale blue. An ideal candidate for a sunny sink or trough.
***F.g.* 'Seeigel' ('Sea Urchin')** Green hair-fine leaves.

F.g. 'Silbersee' ('Silver Sea') Makes a neat, tight little mound to a maximum of 12 cm (5 in), selected both for its dwarfness and its pale silvery blue colouring.

Festuca mairei   Gramineae
L 60 × 60 cm (2 × 2 ft)   *Fl* 1 m (3¼ ft)
Maire's fescue. From a different mould altogether. Longer leaves, shiny grey-green, almost silvery. Spring flower heads, thin and nothing special. If planting in groups, separate the individual specimens, and surround them perhaps with the white-variegated holcus. Average soil in sun.

Festuca paniculata   Gramineae
L 40 × 40 cm (16 × 16 in)   *Fl* 1 m (3¼ ft)
Broader leaved than any of the other fescues described here, smooth, mid-green, paler on the

reverse. Tidy inflorescence of quite large spikelets. Most soils in sun or light shade.

Festuca punctoria   Gramineae
L 15 × 30 cm (6 × 12 in)   *Fl* 30 cm (1 ft)
Porcupine grass. Rigid, sharply pointed leaves, a fine silvery blue. As the leaves lengthen they tend to splay out leaving the centre bare and rather unsightly. Lifting and replanting a little deeper in early summer will keep it neat. A sunny, well-drained spot is important.

Festuca vivipara   Gramineae
L 15 × 30 cm (6 × 12 in)   *Fl* 35 cm (14 in)
Another of those oddities which surprise you by producing tiny plantlets instead of true flowers. Intriguing, certainly, and worthy of garden space as a conversation piece at least. Hair-fine, somewhat grey-green leaves. Each plantlet is carried on a short reddish stalk.

GLYCERIA

Glyceria maxima 'Variegata'   Gramineae
L 60 × 45 cm (2 × 1½ ft)   *Fl* 90 cm (3 ft)
Marvellous cream foliage, only narrowly green-striped and with a pink flush upon emerging in spring. Spread will be determined significantly by soil conditions. Grown literally in water, or in wet ground it may be invasive, but in ordinary border conditions it is more restrained. Open, yellow-green flowering panicles. Cream and blue mix beautifully, so try with the prostrate conifer *Juniperus squamata* 'Blue Carpet', or with blue hostas or hebes.

HAKONECHLOA

Hakonechloa macra 'Aureola'   Gramineae
L 20 × 40 cm (8 × 16 in)   *Fl* 25 cm (10 in)
A choice grass which requires good soil in sun or part shade. A slow starter but gradually forms a mound of

Fig. 20 In a sunny, well-drained position the rigid, silvery blue, pointed leaves of *Festuca punctoria* contrast beautifully with the pink and white variegation and pale pink flowers of *Sedum spurium* 'Variegatum'.

arching, overlapping leaves, bright yellow with rich green stripes, often with a reddish flush towards the autumn. Insignificant flowers, late summer. Makes a classic combination with *Hosta* 'Halcyon'.

HELICTOTRICHON

Helictotrichon sempervirens Gramineae
L 45×75 cm (1½×2½ ft) *Fl* 1.2 m (4 ft)
A fine evergreen grass for a sunny position in ordinary to light soil. Steely grey-blue leaves from a tight central clump form a hemisphere of some 38 cm (15 in) radius. Graceful nodding flower heads on arching stems, spring and early summer. Give it a specimen position among low silvers, purples and pinks.

HOLCUS

Holcus mollis 'Albovariegatus' Gramineae
L 15×40 cm (6×16 in) *Fl* 30 cm (1 ft)
Excellent carpeting grass of bright creamy white effect, with only a narrow central green stripe. Plant under darker upright grasses, or allow to intermingle with the purple-leaved clover, *Trifolium repens* 'Quadrifolium Purpurascens', for an interesting effect. Pale green inflorescences during the summer are not abundant and are best removed before seeding. Average soils in sun or moderate shade.

HORDEUM

Hordeum jubatum Gramineae
L 50×30 cm (20×12 in) *Fl* 75 cm (2½ ft)
Foxtail barley. Highly decorative in the garden or for drying and dyeing. A host of long bristles (awns) arises from the nodding flower spike in early and mid summer, pale green with a hint of red, turning buff. Cut back if untidy in late summer. Best treated as an annual sown direct into the soil in sun in spring or autumn.

HYSTRIX

Hystrix patula Gramineae
L 90×45 cm (3×1½ ft) *Fl* 1.2 m (4 ft)
Bottle brush grass. Aptly named, as the Latin generic name, *Hystrix*, means porcupine. Upright culms from a loose clump of broad green leaves. Long, narrow, erect heads of awned (bristled), regularly spaced spikelets are held initially at an angle, then at right angles to the stem. The individual spikelets are green, sometimes pink-tinged, appearing around mid summer and remaining of interest into the autumn. Good for cutting and drying. Ordinary soil in sun or part shade.

IMPERATA

Imperata cylindrica 'Rubra' Gramineae
L 35×20 cm (14×8 in)
Japanese blood grass. Mid-green leaves, appearing quite late, quickly become blood red at the tips. By autumn the whole plant is aglow – quite astonishing! Prefers good soil which does not dry out and a little shade. Not reliably hardy in extreme cold so be prepared to give protective covering. Christopher Lloyd suggests planting with the black-leaved *Ophiopogon planiscapus* 'Nigrescens'. I would add some bright yellow.

INDOCALAMUS

Indocalamus tessellatus Gramineae
(Bambuseae)
L 1.5×1.8 m (5×6 ft)
A bamboo grown for its enormous shiny leaves achieving a length of almost 60 cm (2 ft) and 10 cm (4 in) wide, weighing down the slender green culms, and bestowing a luxuriant effect. A stand of this will be impressive but invasive.

Evening sunlight filters through the tall *Miscanthus sinensis* 'Variegatus', *Cotinus coggygria* 'Royal Purple' and a foreground clump of *Artemisia* 'Powis Castle'.

The charming *Molinia caerulea* 'Variegata' brightens a
border in dappled shade.

JUNCUS

Juncus concinnus Juncaceae
L 25 × 15 cm (10 × 6 in) *Fl* 30 cm (1 ft)

An interesting rush from Pakistan with erect cylindrical leaves, deep red at the base and nodes, and rather scabious-like, creamy white flower heads. Moist soil in sun.

Juncus decipiens 'Curly Wurly' Juncaceae
L 10 × 30 cm (4 × 12 in) *Fl* 10 cm (4 in)

A much more refined plant in every way than the more familiar corkscrew rush (below). Narrow, cylindrical leaves like coils of the green plastic-coated wire used in the garden. Small brown flower heads in summer. Good to damp soil in sun or part shade.

Juncus effusus 'Spiralis' Juncaceae
L 45 × 90 cm (1½ × 3 ft) *Fl* 45 cm (1½ ft)

Corkscrew rush. The common name is precisely descriptive of the strange mode of growth. The spiralled, cylindrical, shiny green leaves either lie flat on the ground or grow more erect. Pale brown flower heads through the summer. May be grown in damp soil or in water, in sun or part shade.

Juncus inflexus 'Afro' Juncaceae
L 45 × 90 cm (1½ × 3 ft) *Fl* 45 cm (1½ ft)

If you have neither damp soil nor water but have a hankering for a corkscrew rush choose this variety, with blue-green leaves, which will grow in ordinary soil in sun or part shade. Again, pale brown flower heads in summer.

Juncus xiphioides Juncaceae
L 30 × 30 cm (1 × 1 ft) *Fl* 40 cm (16 in)

Couldn't be more different to the foregoing. Erect, iris-like leaves, one margin of which appears papery and colourless, giving the impression of variegation. Pale, flat flower heads from early summer turn a rich reddish colour as the seeds ripen. Forms a loose, spreading clump. Good to moist soil or plant actually in water.

KOELERIA

Koeleria cristata glauca Gramineae
L 15 × 15 cm (6 × 6 in) *Fl* 35 cm (14 in)

Crested hair grass. Of similar stature to the blue fescues but with broader, grey-green leaves and more showy inflorescences – a longer, dense panicle (in early to mid summer) of shiny spikelets arranged in noticeably regular fashion up the flowering stem. Plant at the front of the border, singly or in groups, perhaps with white-variegated plants. A chalk lover but will grow happily in most soils in sun.

Koeleria vallesiana is a very similar plant, confined in Great Britain to limestone hills in Somerset.

LAGURUS

Lagurus ovatus Gramineae
L 30 × 30 (1 × 1 ft) *Fl* 50 cm (20 in)

Hare's tail. Easily grown annual, much used for dried arrangements and dyes well. Its soft, hairy, dense heads, oval-shaped and palest green, becoming creamy white in colour, are also of value in the garden. Sow seed in the chosen spot in spring. Average to light soil in sun.

L.o. nanus is a delightful miniature form.

LUZULA

Luzula alopecurus Juncaceae
L 15 × 15 cm (6 × 6 in) *Fl* 30 cm (1 ft)

A species of woodrush with the emerging brown flower heads enveloped in cocoon-like silky hairs.

Luzula × borrerii 'Botany Bay' Juncaceae
L 15 × 15 cm (6 × 6 in) *Fl* 25 cm (10 in)

New leaves in spring are striped white and sometimes

pink-flushed, a somewhat fleeting phenomenon, often requiring close inspection to be observed at all. The effect is enhanced in shade. The older leaves are a quietly pleasing pale matt green. Typical heads of brown flowers in summer. Damp or dry soil in shade.

Luzula nivea Juncaceae
L 30 × 25 cm (12 × 10 in) *Fl* 50 cm (20 in)

Snowy woodrush. Distinctive, shiny white, compact flowering heads well above the dense clump of deep green evergreen leaves endowed with numerous marginal hairs. A pleasant antidote to brighter colours in sun or shade.

Luzula pumila Juncaceae
L 5 × 8 cm (2 × 3 in) *Fl* 12 cm (5 in)

A tight little mound of narrow, rather stiff leaves, ideal for a sink or trough, or for small pockets in the rock garden. Small brown flower heads in early summer.

Luzula sylvatica forma Juncaceae
L 35 × 30 cm (14 × 12 in) *Fl* 60 cm (2 ft)

This is an interesting form with new shoots almost white, gradually turning green. All these varieties of the greater woodrush are valuable evergreens, providing the densest of slowly spreading ground cover in sun or, perhaps more usefully, in shade, damp or dry. They are broad leaved, slightly hairy at the margins and bear open heads of chestnut brown flowers in mid to late spring.

L.s. 'Hohe Tatra' Sometimes distributed as *L.s.* 'Aurea', this is of very recent introduction and is surely set to be a winner. Brightest yellow winter colour of any of the plants considered in this book, broad leaves contributing to the considerable impact. As the new growth emerges the colouring moves through lime-yellow to fresh yellow-green during the summer, before returning to its brightest best in winter cold.

L.s. 'Marginata' The most commonly encountered variety and an excellent plant. Shiny rich green leaves have neat, narrow, cream margins.

L.s. 'Select' A new, larger and more robust form, with foliage to about 60 cm (2 ft).

L.s. 'Tauernpass' By contrast, a lower-growing variety with slightly broader leaves, reaching only 25 cm (10 in).

Luzula ulophylla Juncaceae
L 10 × 12 cm (4 × 5 in) *Fl* 15 cm (6 in)

Recently introduced to a wider market from New Zealand seed by Graham Hutchins. Most interesting species, with very dark green leaves so V-shaped in section that the abundant silver hairs on the reverse are a prominent feature, giving a splendid silvery effect. The very tight clump sends up flowering stems in early summer topped by a single black spike.

MELICA

Melica altissima 'Atropurpurea' Gramineae
L 90 × 45 cm (3 × 1½ ft) *Fl* 1.2 m (4 ft)

Each spikelet, hanging from the stem at a 45-degree angle, is purple in colour, turning paler as the seed heads mature. Avoid the heaviest soils, otherwise easy in sun or part shade. This, and *M.a.* 'Alba' with almost white spikelets, can be of rather untidy habit but are worth growing for their flower heads.

Melica nutans Gramineae
L 22 × 15 cm (9 × 6 in) *Fl* 30 cm (1 ft)

Nodding or mountain melick. Loose clump of fresh green leaves. Flower heads, late spring and through the summer, hanging as a graceful one-sided panicle of sparse, small, oval-shaped spikelets. Most at home in part shade but full sun is acceptable in good soil.

Melica uniflora 'Variegata' Gramineae
L 20 × 30 cm (8 × 12 in) *Fl* 25 cm (10 in)

A lovely variety of the wood melick with centrally

Although not always hardy *Pennisetum villosum* is well worth raising from seed for its fluffy late summer flower heads.

Phalaris arundinacea 'Feesey's Form' is a less commonly encountered variety of the familiar gardener's garters, with a white central stripe.

white-striped leaves often flushed purple. Lower section of the culms purplish. Good soil in some shade. Plant in groups for best impact. Small dark, beadlike spikelets resemble blackfly on first appearing!

M.u.* var. *albida Bears small white flowers on erect, branched stems over fresh green leaves in late spring and early summer.

MILIUM

***Milium effusum* 'Aureum'**   Gramineae
L 30 × 20 cm (12 × 8 in)   *Fl* 60 cm (2 ft)
Bowles' golden grass. An excellent lightener of part-shaded spots. Every part of this plant is bright yellow: the broad, soft, somewhat floppy leaves, the flowering stems, and the open panicle of tiny spikelets on hair-fine branches in early summer. Slightly more green as the season progresses or in too much shade. Average to good soil in semi-shade. When well suited will self-seed happily. Splendid between green or yellow-variegated shrubs.

MISCANTHUS

Miscanthus floridulus   Gramineae
L 2.7 × 1.5 m (9 × 5 ft)   *Fl* 3 m (10 ft)
Apparently this is the correct name for the grass that has been distributed for years as *M. sacchariflorus*, which is shorter, earlier flowering, and possibly less hardy. To what extent *M. sacchariflorus* is grown at all in our cooler climate it will be interesting to ascertain.

Strong grower, forming erect, slowly spreading clumps. Broad grey-green leaves, 75 cm (2½ ft) in length, with a silvery midrib, arch outwards from the upward thrusting culms. Whitish flower panicles rarely produced except in warmer areas. Use as an isolated specimen or accent plant, as a background plant to lower planting, or as a screen which will reach its full height only in late summer. The dead foliage rustles and flutters but is best cut back by early winter or it will be scattered around the garden by winter winds. Tolerant of most soil conditions, in sun or part shade, but will reward planting in good soil.

Miscanthus sinensis   Gramineae
L 1.8 × 1.5 m (6 × 5 ft)   *Fl* 2.1 m (7 ft)
To date at least 40 forms and cultivars of this wonderfully stately plant have become available, and as there are some quite small varieties this is surely one grass that no garden should be without. Easy-going as to requirements and, while doing best in good moisture-retentive soil, will not object to anything from light to heavy soil in sun or light shade. Erect habit of growth with long leaves cascading from stout culms. One or two forms do not flower regularly, but most do, appearing between mid summer and mid autumn and lasting through the winter. The panicles are roughly fan-shaped with multiple finger-like spikes, shining, silvery pinkish to reddish brown. Selected forms have whiter or redder colouring. Best planted in some isolation or with only low-growing plants around them so that their architectural qualities may be fully appreciated. Flowering stems may be cut and dried for indoor decoration.

***M.s.* 'Cabaret'**   A wonderful new variegated form with green margins and stripes to a white leaf.

***M.s.* 'Flamingo'**   Pale purplish pink flowers appear in late summer, 1.8 m (6 ft) tall, over 1.2 m (4 ft) foliage.

***M.s.* 'Goldfeder'**   This has yellow margins to the leaves.

***M.s.* 'Goliath'**   A strong grower with larger flower heads in early autumn, 2.5 m (8 ft).

***M.s.* 'Gracillimus'**   Maiden grass. Does not regularly produce flowers in cooler climates, but it has a pleasing form with very narrow arching leaves in a dense, compact clump providing interest right through the winter. 1.2 m (4 ft) foliage.

M.s. 'Graziella' An excellent narrow-leaved variety, similar to 'Gracillimus', but a good flowerer, pale, silvery panicles appearing in late summer, reaching 1.8 m (6 ft).

M.s. 'Kascade' Wide-open, lax panicles, rich deep red at first, and reaches 2.1 m (7 ft).

M.s. 'Kleine Fontane' A smaller variety with long, narrow leaves and pinkish flower heads, quite early, from mid to late summer onwards, 1.5 m (5 ft) over 1 m ($3\frac{1}{4}$ ft) foliage.

M.s. 'Malapartus' The broad leaves attain a purplish flush by autumn. The flowers, to 1.6 m ($5\frac{1}{2}$ ft), opening purple-red, become silver from late summer.

M.s. 'Morning Light' This has very narrow foliage to 1.2 m (4 ft), each leaf having a thin white marginal stripe in addition to the silvery midrib.

M.s. 'Nippon' With its 90 cm (3 ft) clump of narrow leaves, this is an excellent variety for the smaller garden. Early autumn flower heads to 1.2 m (4 ft) open a good red.

M.s. 'Punktchen' With the yellow cross-banding of the more familiar 'Zebrinus' but on a much smaller plant, reaching 1 m ($3\frac{1}{4}$ ft).

M.s. *purpurascens* Flowers, when they do appear, which is not regularly, reach 1.2 m (4 ft) in mid autumn, but this form compensates with foliage colour – a deep purplish flush becomes apparent in the summer, the normally silver median stripe being pink, and intensifies to bright brown and reddish tones. The foliage is 75 cm ($2\frac{1}{2}$ ft) high.

M.s. 'Rotsilber' Prominent central silvery stripe to the narrow leaves. Good red flower heads, especially on opening out in early autumn, 1.5 m (5 ft) over 1 m ($3\frac{1}{4}$ ft) foliage.

M.s. 'Silberfeder' ('Silver Feather') Similar to the species in size, it develops excellent silvery flower heads in early autumn, becoming pale pinkish brown, and lasting well into the winter.

M.s. 'Silberspinne' Forms an erect 1 m ($3\frac{1}{4}$ ft) clump of narrow leaves, with 1.2 m (4 ft) panicles of long, open, spidery fingers from early autumn, reddish at first, becoming silvery.

M.s. 'Sirene' Another tallish variety, the 1.8 m (6 ft) flower heads opening rich red-brown in early autumn.

M.s. 'Strictus' One of the few grasses with transverse yellow bars across the leaves, appearing from mid summer on. Pinky brown flower heads rise to 1.8 m (6 ft) in the autumn. Of particularly erect and narrow habit.

M.s. 'Variegatus' A truly stately variegated grass – a veritable fountain of markedly white-striped leaves. Use as a stunning contrast to purple-leaved and dark green shrubs, or as a harmonizing tone but sharply contrasting form in paler planting schemes. Not reliably free-flowering, but when they do appear in mid autumn the panicles are pinkish, to 1.8 m (6 ft) over 1.5 m (5 ft) foliage.

M.s. 'Zebrinus' With similar transverse banding to M.s. 'Strictus' but is of more spreading habit. The two varieties are probably totally confused in the trade.

Miscanthus tinctoria 'Nana Variegata'
Gramineae
L 30 × 45 cm (12 × 18 in)   *Fl* 50 cm (20 in)

Here is a super little plant which should soon be getting around in the trade. The smallest of the family, small enough for the tiniest garden. Forms a loose, spreading clump of fresh, yellow-green leaves with narrow white stripes. Pale brownish flower heads in early autumn. Good soil in sun or part shade.

Miscanthus yakushimensis Gramineae
L 75 × 45 cm ($2\frac{1}{2}$ × $1\frac{1}{2}$ ft)   *Fl* 1.2 m (4 ft)

We conclude this splendid genus with another dwarf species, much like a small *M. sinensis* but with pale leaves, and pale pinky brown flowers with fewer, long, slender fingers.

MOLINIA

Molinia caerulea 'Heidebraut' ('Heather Bride')
Gramineae
L 60 × 40 cm (24 × 16 in)   *Fl* 1.2 m (4 ft)
The purple moor grass is another species which has benefited from the careful selection of distinctive varieties by German nurserymen. In the wild it grows in areas of damp, acidic moorland but, in fact, is quite at home in most average to moist soils in sun or part shade. This variety has yellowish stems and inflorescences with glistening spikelets and seed heads.

M.c. 'Moorhexe' ('Bog Witch') Less tall, with foliage to about 35 cm (14 in), of rigidly erect form. Tight, almost needle-like, dark flower heads stand to attention in late summer, reaching 45 cm (1½ ft). The upright growth of this form suits it admirably for use as a vertical specimen among lower, light-toned plants.

M.c. 'Variegata' Neat and attractive in every respect. From a compact 35 cm (14 in) clump arise leaves of cream with pale green stripes which remain attractive long after they have faded to a pale buff colour in the autumn. Creamy yellow stems bear nicely contrasting purplish spikelets rising to about 60 cm (2 ft) in late summer. Avoid the hottest spots and plant with any blue to grey-green foliage.

Molinia caerulea arundinacea 'Bergfreund'
Gramineae
L 75 × 60 cm (2½ × 2 ft)   *Fl* 1.6 m (5½ ft)
The English translation of the varietal name is 'Mountain's Friend'. This and the following are varieties of a taller-growing subspecies, and make quite imposing plants for many weeks after the inflorescences appear in late summer. The foliage in this form provides marvellous golden yellow autumn colour and the small spikelets give a particularly delicate effect.

M.c.a. 'Karl Foerster' Fairly erect and with open, airy, purple flower heads to about 1.4 m (4½ ft).

M.c.a. 'Skyracer' Features the tallest flower heads of all, reaching some 2.2 m (7½ ft).

M.c.a. 'Windspiel' ('Windplay') The 2 m (6½ ft) yellowish inflorescences are more substantial, and have larger spikelets. The name is indicative of its graceful swaying in the breeze.

M.c.a. 'Zuneigung' ('Affection') Of similar height; it derives its name from the habit of its culms of entwining and disengaging in the wind – apparently reminiscent of an amorous couple!

MUHLEMBERGIA

Muhlembergia japonica 'Cream Delight'
Gramineae
L 20 × 90 cm (8 × 36 in)   *Fl* 20 cm (8 in)
Brought to me by two Japanese variegated plant enthusiasts, the naming to be confirmed. A lovely plant, creamy white margined and striped, purplish at the nodes, and of distinctive habit: the culms are initially erect, but as they lengthen soon become prostrate, splaying out to fill a circle of some 90 cm (3 ft) diameter. Flower heads at the ends of these stems, dense clusters of silver and purple spikelets, in late summer. Proving thoroughly hardy. Good soil in sun or part shade.

PANICUM

Panicum clandestinum   Gramineae
L 50 × 25 cm (20 × 10 in)   *Fl* 75 cm (2½ ft)
Deer tongue grass. Will require protection in winter away from warm areas or micro-climates. Interesting leaves, in excess of 2.5 cm (1 in) wide but only 17 cm (7 in) long, with rounded, hairy base clasping the stems. Flower heads late summer and early autumn, typically diffuse and airy. Good soil in sun.

Panicum miliaceum   Gramineae
L 80 × 35 cm (32 × 14 in)   *Fl* 1 m (3¼ ft)
An important food plant in many parts of the world, millet may be grown in the garden, and its culms, with their dense flopping inflorescences, cut for fresh or dried arrangements. Sow seed in spring in good soil in sun. ***P.m.* 'Violaceum'** has purple flower heads.

***Panicum virgatum* 'Haense Herms'**   Gramineae
L 60 × 30 cm (2 × 1 ft)   *Fl* 90 cm (3 ft)
Switch grass offers two good features in all its cultivars – a positively vertical line, and late summer to autumn flower heads which, although large, are comprised of such tiny spikelets, so well spaced, that they give a wonderfully hazy effect. This and the varieties **'Rehbraun'**, **'Rotstrahlbusch'** and **'Rubrum'** all feature reddish brown foliage during the course of the season, intensifying in the autumn. Any soil that is not too heavy, in sun or light shade.
***P.v.* 'Strictum'**   A narrowly erect form growing to about 1.4 m (4½ ft) in flower, with green leaves becoming bright golden yellow in autumn, then fading paler. May be enjoyed until late winter before cutting back.

PENNISETUM

Pennisetum alopecuroides   Gramineae
L 90 × 75 cm (3 × 2½ ft)   *Fl* 1 m (3¼ ft)
Swamp foxtail or fountain grass. The pennisetums as a genus generally feature bristly flowers resembling hairy caterpillars or bottle brushes. *P. alopecuroides* grows in dense clumps of bright green, narrow, arching leaves. In late summer and early autumn the long spikes appear, varying from a bluish purple to pale pinkish brown with, for the brief duration of their

The wonderful bright colouring of the bamboo *Pleioblastus viridistriatus.*

appearance, bright orange-red stamens. Good soil in full sun. Protect from extreme cold.

P.a. 'Hameln' A low-growing form, to about 50 cm (20 in), and flowers earlier.

P.a. 'Woodside' A free-flowering, colourful form.

Pennisetum macrourum Gramineae
L 100×75 cm (3¼×2½ ft)   Fl 1.5 m (5 ft)
Normally hardy, this is a taller species from southern Africa with narrow, 30 cm (1 ft), cylindrical spikes in late summer, green becoming pale brown. Decorative in the garden and for flower arrangements. Well-drained soil in sun.

Pennisetum orientale Gramineae
L 30×45 cm (1×1½ ft)   Fl 45 cm (1½ ft)
One of the best grasses for floral effect, and for its long period of interest, but requiring winter protection in colder areas – or it may be raised from seed sown in spring under glass. The grey-brown leaves form a dense clump surmounted by numerous pinkish bottle brushes, long and feathery, from mid summer on into the autumn. Sunny, well-drained soil.

Pennisetum setaceum Gramineae
L 45×60 cm (1½×2 ft)   Fl 75 cm (2½ ft)
Not very hardy and best grown as an annual in most areas, starting the seed off under glass in spring. Graceful, arching habit. Long, pink, feathery spikes over a long period from mid summer to early autumn. Ordinary soil in sun or part shade.

Pennisetum villosum Gramineae
L 45×60 cm (1½×2 ft)   Fl 60 cm (2 ft)
Feather top. Best grown as an annual from seed sown under glass in spring, as it is not always hardy. The plump, feathery flower heads, slightly pinkish in colour, are borne in late summer and early autumn, and are useful for cutting, drying and dyeing. Good soil in sun or part shade.

PHALARIS

Phalaris arundinacea 'Feesey's Form'
Gramineae
L 60×60 cm (2×2 ft)   Fl 90 cm (3 ft)
It is another variety of *P. arundinacea* (see 'Picta', below) which affords most gardeners their introduction to ornamental grasses. The experience is almost invariably an unhappy one as they battle with its invasive nature. The woman in Chapter 1 was undoubtedly a case in point! However, these are all strikingly handsome plants and eminently worthy of use in the right place. Use among established shrubs or where cover is needed. If they lose some brightness around mid summer cut them back to about 20 cm (8 in) for a second flush of growth. Will tolerate dry soils or, at the other extreme, will grow in water, in sun or shade.

'Feesey's Form' is the best and least commonly seen variety. Wonderful white effect, with only narrow green margins and stripes. A little less tall and vigorous than the other forms.

P.a. luteopicta Another most attractive form with pale creamy yellow stripes. Benefits from a mid summer cut-back. Flower heads 1.2 m (4 ft) tall in mid summer over 75 cm (2½ ft) foliage.

P.a. 'Picta' Already mentioned, this is a markedly white-striped form, and is fast-spreading.

P.a. 'Tricolor' The white margins and stripes manifest a pronounced purplish pink flush, maintained throughout the season, especially in sun.

Phalaris canariensis Gramineae
L 60×30 cm (2×1 ft)   Fl 75 cm (2½ ft)
Canary grass. Used for dried flower arrangements and for bird seed. Dense, short, fat, pale green heads from early summer to early autumn. Annual, grown from seed sown in spring where it is to flower. Any average soil in sun or part shade.

PHLEUM

Phleum pratense   Gramineae
L 90 × 60 cm (3 × 2 ft)   *Fl* 1.2 m (4 ft)
Timothy grass, cat's tail. Native British grass which may be grown in the garden for its long, narrow, cylindrical flower spikes which can be cut for fresh or dried arrangements, or for dyeing. Most soils in sun or part shade.

PHRAGMITES

Phragmites australis giganteus   Gramineae
L 3 × 1 m (10 × 3¼ ft)   *Fl* 3.5 m (12 ft)
This form is even taller than the British native Norfolk reed, vigorous, indeed invasive, and should be contained if planted anywhere other than where its rapid spread is really desired, such as in lakeside settings. Warm locations are preferred, or submerged in water.
***P.a.* 'Variegatus'**   The best taller yellow-variegated grass, 1.4 m (4½ ft) with richly coloured stripes – but invasive, so will need containing or watching closely. Doesn't have to be grown in water – good soil is perfectly adequate. Splendid shaggy purple flower heads 1.5 m (5 ft) high in late summer. A form with creamy white variegation is also sometimes seen under the same name. Infinitely better is a newer variety called **'Karka'** which is shorter and with far more conspicuous creamy white stripes.

PHYLLOSTACHYS

Phyllostachys aurea   Gramineae (Bambuseae)
L 3.6 × 2 m (12 × 6½ ft)
This bamboo does not really live up to the promise of its botanical name, the culms never being more than yellowish green. That said, the plant itself is always pleasing and displays the odd characteristic of increasingly swollen and distorted lower nodes. For much

brighter yellow culms look for the variety called **'Holochrysa'**. Best in sun. Moderate spreader.

Phyllostachys nigra   Gramineae (Bambuseae)
L 4 × 2 m (13 × 6½ ft)
Black bamboo. Thick canes become shiny black in their third season. The abundant foliage and arching nature of the moderately spreading canes add grace. Sun or part shade.

PLEIOBLASTUS

Pleioblastus humilis pumilis   Gramineae (Bambuseae)
L 1 × 1 m (3¼ × 3¼ ft)
A short but invasive bamboo with bright green leaves and very slender culms. Grow it in sun or shade, either contained, or where it can be mown around, or where it really doesn't matter if it spreads as dense, low cover. Bamboos such as this are useful in that, once established, they will colonize difficult spots such as dry shade under trees.

Pleioblastus pygmaeus   Gramineae (Bambuseae)
L 20 × 75 cm (8 × 30 in)
A real dwarf, but a real spreader. With its small leaves it is by no means unattractive and there are situations where its invasive nature may be appreciated, but a physical barrier will usually be necessary in smaller settings. Sun or shade.

Pleioblastus variegatus   Gramineae (Bambuseae)
L 100 × 45 cm (3¼ × 1½ ft)
Striking creamy white variegation. Only a moderate spreader, forming dense clumps of erect, very slender culms. Excellent in good soil in sun or part shade, between or against darker-leaved shrubs to emphasize its variegated leaves.

The withdrawal of pigment from the leaf margins of the bamboo *Sasa veitchii* gives the impression of variegation.

**The arching leaves of *Spartina pectinata* 'Aureamarginata'
demonstrate a grace of form.**

Pleioblastus viridistriatus   Gramineae
(Bambuseae)
L 100 × 45 cm (3¼ × 1½ ft)
A magnificently bright bamboo, leaves predominantly yellow with variable stripes. Best colouring in sun and in good soil. In a sheltered spot where the foliage suffers little winter damage the plant may be left to increase gradually in height, but the best policy is to cut it down to the ground in early spring for an annual flush of fresh new growth.

POA

Poa acicularifolia   Gramineae
L 4 × 15 cm (1½ × 6 in)   *Fl* 10 cm (4 in)
A fascinating little mat-forming grass. Tiny leaf blades alternating up the short erect stems are surprisingly stiff and quite prickly. Rather sparse flower heads on delicate wiry stems in late spring. For lighter soils in sun.

Poa buchananii   Gramineae
L 10 × 15 cm (4 × 6 in)   *Fl* 20 cm (8 in)
Another small, distinctive grass, forming tight clumps of deep grey-blue, purple-tinged leaves which are flat to slightly boat-shaped, fairly stiff and held in a more or less horizontal plane. Flower spikes emerge in early summer in a colour exactly matching that of the leaves, becoming olive-grey. Well drained soil in sun.

Poa chaixii   Gramineae
L 38 × 38 cm (15 × 15 in)   *Fl* 90 cm (3 ft)
Broad-leaved meadow grass. Useful, robust plant with broad, shiny evergreen leaves of rich green, making a solid clump. Numerous flowering stems bear clouds of purplish spikelets in late spring and early summer, standing erect, well above the foliage. A splendid grass for shade, although it will happily tolerate full sun so long as the soil is not too dry.

Poa colensoi   Gramineae
L 20 × 20 cm (8 × 8 in)   *Fl* 40 cm (16 in)
In considerable contrast, this hardy New Zealand meadow grass is very narrow-leaved, cylindrical and of excellent blue, not unlike the better varieties of *Festuca glauca*, but perhaps a richer colour. Evergreen. Summer flowers.

Poa × jemtlandica   Gramineae
L 10 × 15 cm (4 × 6 in)   *Fl* 25 cm (10 in)
When I obtained this grass it was described simply as 'a rare hybrid in the wild, collected Ben Nevis'. I was surprised at flowering time in early summer to discover that this is another viviparous grass, bearing tiny red-based plantlets on amazingly tough stems in the place of true flowers and seed. Forms a small mound of grey-green foliage. Average soil in sun.

Poa labillardieri   Gramineae
L 60 × 90 cm (2 × 3 ft)   *Fl* 90 cm (3 ft)
Narrow, blue-grey evergreen leaves arch upwards and outwards from a central clump. For a specimen position, preferably with adjacent silver foliage. Summer inflorescence rather open and sparse. From Australia and New Zealand, it needs a sunny, well-drained spot, and may not be hardy in cold climates.

POLYPOGON

Polypogon monspeliensis   Gramineae
L 50 × 35 cm (20 × 14 in)   *Fl* 75 cm (2½ ft)
Annual beard grass. Many dense, cylindrical, silky flower spikes up to 15 cm (6 in) long are produced throughout the summer months. Sow *in situ* in late spring in ordinary to good soil in sun.

PSEUDOSASA

Pseudosasa japonica   Gramineae (Bambuseae)
L 4 × 2 m (13 × 6½ ft)

A very plain bamboo, adequate for screening but generally best avoided in favour of any of the more ornamental species.

RHYNCHELYTRUM

Rhynchelytrum repens   Gramineae
L 30 × 60 cm (1 × 2 ft)   *Fl* 60 cm (2 ft)
Ruby, or Natal grass. Open heads of small, fluffy rosy red or pinkish flowers on a plant of rather lax habit. Sow seed under glass in spring or give winter protection to this rather tender perennial. Good soil in a sunny, sheltered spot. May be dried and dyed.

SACCHARUM

Saccharum ravennae   Gramineae
L 90 × 90 cm (3 × 3 ft)   *Fl* 1.8 m (6 ft)
Ravenna grass. Preferred in colder zones of the United States to the pampas grass, the situation seems to be reversed in Great Britain, where it is hardy only in warmer localities. Narrow, grey-green arching foliage colours well in the autumn. Flowering plumes in late summer and early autumn, much narrower and more open than those of the pampas grass, one-sided, and silvery to purplish-grey in colour. Good, well-drained soil in sun.

SASA

Sasa veitchii   Gramineae (Bambuseae)
L 1.2 × 1.2 m (4 × 4 ft)
An invasive, broad-leaved bamboo featuring pale edges to the leaves. This is due to a withdrawal of colour from the current year's foliage later in the season, rather than being a true variegation, although this is the overall effect. Grows in sun or shade. See under *Pleioblastus humilis pumilis* and *P. pygmaeus* for comments on the use of invasive bamboos.

SCHOENUS

Shoenus pauciflorus   Cyperaceae
L 25 × 30 cm (10 × 12 in)
A useful sedge featuring rare dark tones. Dense tuft of straight, narrowly cylindrical leaves, green only at the base, thereafter a really deep brownish maroon – a colour which stands out well against pale shingle or chippings, or an underplanting of low, light-toned foliage. The sparse flowers are barely apparent at the tips of the culms. Any soil in sun.

SCIRPUS

Scirpus lacustris 'Albescens'   Cyperaceae
L 120 × 45 cm (4 × 1½ ft)   *Fl* 1.2 m (4 ft)
This form of the bulrush gives an interesting effect to the larger pool, with tall, narrow, cylindrical stems rising erect and virtually leafless from the water, white to all intents and purposes, though actually narrowly green-striped. Branched flower head of reddish brown spikelets. Plant up to 15 cm (6 in) below the surface of the water.

Scirpus lacustris tabernaemontani 'Zebrinus'
Cyperaceae
L 100 × 45 cm (3¼ × 1½ ft)   *Fl* 90 cm (3 ft)
Transversely banded leaves, creamy white alternating with grey-green. A most showy effect. Similar treatment to the previous variety.

SESLERIA

Sesleria caerulea   Gramineae
L 15 × 25 cm (6 × 10 in)   *Fl* 25 cm (10 in)
Blue moor grass. Grows wild in northern parts of Britain, and is well worth cultivating in the garden. Very dense tuft. Leaves a good grey-blue above and shiny dark green beneath, carried in a horizontal fashion which results in the blue predominating. The

**The feathery heads of *Stipa calamagrostis* appear from mid
summer right into the autumn.**

The showy glistening flower heads of *Stipa gigantea* are here picked out by the sun against a background clump of *Miscanthus floridulus*.

flower heads, appearing quite early, around mid spring, are short and rounded, violet to silver-grey. Medium to light soils in sun or part shade.

Sesleria heufleriana   Gramineae
L 35 × 45 cm (14 × 18 in)   *Fl* 60 cm (2 ft)

Fresh green leaves are topped by narrow purplish spikes throughout the summer.

Sesleria nitida   Gramineae
L 30 × 40 cm (12 × 16 in)   *Fl* 50 cm (20 in)

Not exciting, but certainly pleasing. Leaves long, very smooth, pale grey-green, each terminating in a tiny spike. Narrow, cigar-shaped flower spikes in late spring also pale greyish. Any average soil in sun or part shade. Excellent in paler schemes with silver and white variegation.

SETARIA

Setaria italica   Gramineae
L 60 × 30 cm (2 × 1 ft)   *Fl* 60 cm (2 ft)

Foxtail millet. Popular seed for cage birds, and much used for cutting and drying. Long, heavy flower heads in late summer and early autumn. Annual. Sow seed in spring where it is to flower.

Setaria lutescens   Gramineae
L 60 × 30 cm (2 × 1 ft)   *Fl* 60 cm (2 ft)

Yellow bristle grass. Grey-green foliage, and yellow or reddish bristly spikes, densely clustered along the stem. Annual. Treat as above.

SHIBATAEA

Shibataea kumasasa   Gramineae (Bambuseae)
L 1.2 × 1 m (4 × 3¼ ft)

A neat bamboo of only slowly spreading habit, with very slender culms arising earlier in the year than most. Sun or shade.

SINARUNDINARIA

Sinarundinaria murielae   Gramineae (Bambuseae)
L 4 × 2 m (13 × 6½ ft)

A most graceful specimen bamboo for good to slightly damp soil, preferably with some shelter. Almost tree-like in shape, with initially erect culms becoming arching, almost sagging, and well clothed with narrow, 12 cm (5 in) long leaves.

Sinarundinaria nitida   Gramineae (Bambuseae)
L 4 × 2 m (13 × 6½ ft)

Not dissimilar to the above but with smaller, less bright green leaves, giving an even more delicate effect. The culms are dark purplish. In both species the current year's culms shoot up, quite straight, to 2–3 m (6½–10 ft), not branching at all until the second season. Both may also be used for screening, planted 1.5 m (5 ft) apart.

SORGHASTRUM

Sorghastrum avenaceum   Gramineae
L 75 × 45 cm (2½ × 1½ ft)   *Fl* 1 m (3¼ ft)

Will need some protection in cold areas but well worth growing for its leaves which are greyish, sometimes yellowish green, purple-flushed on hairy purple culms, and for its mid to late summer flower heads. These are quite dense panicles of shiny reddish spikelets with yellow stamens. Prefers fertile, well-drained soil in sun.

SORGHUM

Sorghum halepense   Gramineae
L 120 × 60 cm (4 × 2 ft)   *Fl* 1.8 m (6 ft)

A robust grass of spreading habit for a sunny position in fertile soil, and requiring protection away from mild localities. Open purplish flower heads.

SPARTINA

Spartina pectinata 'Aureamarginata'
Gramineae
L 120 × 90 cm (4 × 3 ft)   *Fl* 1.5 m (5 ft)
Long, ribbon-like leaves arching out and downwards, yellow-green with a yellow marginal band. Forms loose, spreading, but not uncontrollable clumps, especially in wet soils. The specific epithet *pectinata* describes the comb-like nature of the early autumn flowers, although the 'teeth' are held at an angle. Greenish with purple stamens, they make an unusual addition to dried flower arrangements.

SPODIOPOGON

Spodiopogon sibiricum   Gramineae
L 90 × 40 cm (36 × 16 in)   *Fl* 1.5 m (5 ft)
Rarely encountered, but grown for its open, tapering, reddish panicles, the white hairy spikelets borne on ascending branches from mid to late summer. Broad, fresh green leaves, again tapering, held horizontally to the erect culms, and developing purplish autumn colouring. Forms moderately spreading clumps in good soil in sun or part shade.

STENOTAPHRUM

Stenotaphrum secundatum 'Variegatum'
Gramineae
L 10 × 30 cm (4 × 12 in)   *Fl* 15 cm (6 in)
Intriguing and striking, but frost-tender, requiring heated greenhouse or conservatory protection to survive our colder winters. Can be used in pots or in summer bedding schemes. Water well. Short, broad leaves, strongly cream-striped, arise from the smooth, almost rigid, stems which creep along the ground. The peculiar inflorescences are produced in late summer.

STIPA

Stipa arundinacea   Gramineae
L 45 × 90 cm (1½ × 3 ft)   *Fl* 45 cm (1½ ft)
Pheasant grass. Open, airy panicles of shining brown spikelets on very drooping stems. Tight clumps of shiny evergreen leaves, becoming increasingly bronzed as they mature, and acquiring orange and red streaks from late summer and through the winter. Sunny position in good, even heavy soil. Generally hardy except in cold northern areas, but even if the plant is killed seedlings will usually appear.
***S.a.* 'Autumn Tints'** and **'Golden Hue'**   Selected by Graham Hutchins as being particularly good colour forms, the former for its reddish brown colouring and the latter for its pale yellow-green leaves – quite different.

Stipa calamagrostis   Gramineae
L 75 × 120 cm (2½ × 4 ft)   *Fl* 1.2 m (4 ft)
Valued for its long flowering period, from early summer into the autumn. The feathery panicles on arching culms which are reddish at the base, may reach 25 cm (10 in) in length, waving gracefully in the breeze. Leaves, in a dense clump, are bluish green. Easy to grow in most soils in sun.

Stipa gigantea   Gramineae
L 75 × 120 cm (2½ × 4 ft)   *Fl* 2 m (6½ ft)
A glorious specimen plant with erect culms bearing enormous open, oat-like heads of bristled spikelets, around mid summer, over a dense clump of narrow, evergreen foliage. Both culms and flower heads attain a glowing golden hue as they mature. Looks good against a background of darker-leaved plants, but don't crowd it. Average to light soil in sun.

Stipa pennata   Gramineae
L 45 × 35 cm (18 × 14 in)   *Fl* 75 cm (2½ ft)
This species, too, features attractive flower heads but

in a totally different idiom. Though each head bears only a few spikelets it is bestowed with delicate substance by the feathery awns which may reach 25 cm (10 in) in length. Flowers mid to late summer over a tight clump of mildly arching foliage. Ordinary to light soil in sun.

Stipa tenuissima   Gramineae
L 50 × 30 cm (20 × 12 in)   Fl 60 cm (2 ft)
The narrowest of yellow-green foliage grows in particularly erect fashion, topped in mid summer by drooping flower heads of pure gossamer – surely among the most delicate of all grassy inflorescences, resembling billowing clouds when planted in a group, or rolling ocean waves. Ordinary soil in sun.

S. tenacissima   is of no ornamental interest.

TYPHA

Typha angustifolia   Typhaceae
L 90 × 60 cm (3 × 2 ft)   Fl 1.5 m (5 ft)
Lesser reed mace. Not strictly within the limits of this book, but near enough to warrant inclusion. These are for planting in water or very wet soil. The familiar poker heads, the female spikes, which appear below the smaller, short-lived, terminal male spike, are popular for use in dried arrangements. Reddish brown pokers in late summer borne on stiff stems above narrow green leaves. More restrained than the next species but still too vigorous for anything smaller than a large pool. Plant to a depth up to 15 cm (6 in).

Typha latifolia   Typhaceae
L 1.5 × 1 m (5 × 3 ft)   Fl 2.5 m (8 ft)

Great reed mace. Too big and aggressive for the average garden pool, best used for lakeside planting. Leaves broad, grey-green. Dark brown pokers can be up to 30 cm (1 ft) in length, on strong, absolutely erect stems, in late summer.
***T.l.* 'Variegata'**   A splendid new variety, strikingly white-striped. Probably somewhat less vigorous than the type, but still invasive.

Typha minima   Typhaceae
L 45 × 20 cm (18 × 18 in)   Fl 75 cm (2½ ft)
The most sensible species for most of us. More refined in all respects. Narrow, grassy leaves. Flowering stems carry shorter, more rounded heads of reddish brown. Plant in wet soil or in the pool to a depth of 10 cm (4 in) Spreads controllably.

ZEA

Zea mays   Gramineae
L To 150 × 45 cm (5 × 1½ ft)
Maize, Indian corn. Important, widespread food crop. Some highly ornamental forms may be used in the border, as 'dot' plants in summer bedding, or in pots in the greenhouse. Various strains are available, green and white, or with additional yellow and pink, as in **'Harlequin'** and **'Quadricolor'**, or with white, yellow, orange, red, blue or brown kernels, e.g. **'Amoro'** and **'Multicolor'**. Half-hardy annual. Sow seeds early in the year in warmth, planting out after frosts in good soil, keeping well watered. The multi-fingered flower heads at the top of the stems are male, the lower, female inflorescences develop into the edible cobs.

Index

Page number in *italics* indicate illustrations

ACKNOWLEDGEMENTS

The publishers are grateful to the following for granting permission to reproduce the colour photographs: Nigel Taylor (front cover, pp. 2, 6, 15, 22, 30, 43, 46, 54, 55, 59, 62, 63, 66, 74, 75, 83, 87, & 91); Tim Sandall/*The Gardener* (back cover, pp. 18, 35 & 51); John Fielding (pp. 11 27, 39, 67, 71, 78, 79, 86 & 90).